Praise for *The Financial Ti[mes Guide to]* *Business W[riting]*

'Engaging, effective, persuasive writing is an incredibly important and influential business skill, yet one that's often neglected. If you can write well, you can command attention. And you can compel your audience – whether your colleagues, your customers, your prospects or your suppliers.

It's a fascinating subject – and a fascinating book. Without doubt, it's one of the best guides on business writing available today, expertly written and with clear, understandable guidance throughout. It will supercharge your writing and fast-track your business success.'

Kate Allen, Head of UK & Ireland Marketing, BP Castrol

'Every serious business professional should have a copy of this book; it's an absolute godsend. One of the three business books I always have on my desk to refer to, it's worth its weight in gold. In fact it's worth its weight in saffron.

If you want to write better proposals, reports or presentations, get your business recommendations or requests actioned or even shape your company's marketing and advertising, this is the book for you. In short: if you want to get ahead in business, get this book.'

Sheridan Thompson, CRM Director, The Walt Disney Company

'The definitive book on business writing and a really good read to boot. Forget dry, dusty tomes: Atkinson brings the subject to life in a way that makes it as enjoyable as it is illuminating. Five stars.'

Jason Longley, Head of Account & Business Development, ACE European Group

The Financial Times Essential Guide to Business Writing

How to write to engage, persuade and sell

Ian Atkinson

PEARSON

Harlow, England • London • New York • Boston • San Francisco • Toronto • Sydney
Auckland • Singapore • Hong Kong • Tokyo • Seoul • Taipei • New Delhi
Cape Town • São Paulo • Mexico City • Madrid • Amsterdam • Munich • Paris • Milan

Pearson Education Limited

Edinburgh Gate
Harlow CM20 2JE
Tel: +44 (0)1279 623623
Fax: +44 (0)1279 431059
Website: www.pearson.com/uk

First published in Great Britain in 2012

ISBN: 978-0-273-76113-6

British Library Cataloguing-in-Publication Data
A catalogue record for this book is available from the British Library

Library of Congress Cataloging-in-Publication Data
A catalog record for this book is available from the Library of Congress

10 9 8 7 6 5
15

Typeset in Stone Serif 8.75pt by 3
Printed by Ashford Colour Press Ltd., Gosport

Contents

Publisher's acknowledgement

We are grateful to the following for permission to reproduce copyright material:

Text
Extract on page 18 from Apocalypse Japan, *The Sun*, 11 March 2011, The Sun/NI Syndication.

In some instances we have been unable to trace the owners of copyright material, and we would appreciate any information that would enable us to do so.

About the author

Ian Atkinson is a multi award-winning creative director at one of the UK's biggest agency groups.

He's written in every major medium and for some of the best-known brands in their sectors – including Avis, Barclays, Dyson, Macmillan, National Geographic, Oxfam, Sky and Zurich.

He's also a board director who's written, edited or critiqued hundreds of business proposals, plans, presentations, reports, pitch documents and marketing materials.

Ian has a degree in psychology, is a regular conference speaker and runs training days on business writing. He even writes a blog that people actually read.

Introduction

You learn to write when you're what, four or five?

It's that easy.

Which means by the time you start your career, you've had years of practice.

And you want your colleagues to realise just how clever and talented you are.

So you make sure your writing is full of complex ideas. You use a few long words (including the latest terminology) and you sprinkle in a little of your creative genius.

Et voilà! You're a pretty gifted writer.

At least, that's what many people seem to think.

Which is a little odd, really.

Because they know that the ability to turn on a tap doesn't make them a plumber.

And that being able to change a fuse doesn't make them an electrician.

Yet they don't realise that the ability to bash out a few sentences while a red wiggly line underlines your spelling mistakes doesn't make you a convincing wordsmith.

Which is why their presentations fall flat. Why their proposals fail to hit the mark. Why their reports get filed, not actioned. And why their ideas get sidelined, not fast-tracked.

Make no mistake, the ability to write well in business can make a huge difference – to your personal success and your business's success.

In this, *The FT Essential Guide to Business Writing*, we'll explore how it's done.

How to keep your audience's attention – with powerful story-telling, beguiling personality and effective use of the medium.

How to engage your audience with style, structure and substance.

How to win them over with cleverly-thought-out-yet-simply-explained arguments.

And how to persuade them with cunning use of language and psychology.

Throughout, we'll look at examples from business – both good and bad. And we'll look at examples from the world of advertising, to uncover the secrets you can steal from the world's most persuasive writers.

Because, to succeed at the highest level in business, you need those around you to work with you. To be persuaded by you, motivated by you and led by you.

The best business writing can help you do just that.

And this book will help you craft the best business writing.

Ready? Then turn the page and let's get cracking.

part

Planning

1. Powerful business writing: Your objective
2. Proposition: The thin end of your wedge
3. Content: Interrogation & insight
4. Context: Audience & medium
5. Concept: Bringing your story to life
6. Essential Plan checklist

Let me show you something.

It's a five-minute flipchart exercise I sometimes show clients. A simple demonstration of just how easy it can be to make everything you write more pleasing, potent and persuasive.

On a flipchart I write the headline:

> *A few techniques to minimise the shortcomings in writing designed to persuade.*

Then I say, *tip one*: make your writing active and personal. By saying things like 'your writing', for example. Write as one individual talking to another individual.

So, I make a change to the headline on the flipchart. It now reads:

> *A few techniques to minimise the shortcomings in <u>your</u> writing designed to persuade.*

Second tip: use 'hardwired words'. There are words that just seem to grab people's attention, almost as if they're hardwired in our brains. One example is the word 'guaranteed'. Another is 'secrets'. So the headline now becomes:

> *A few secrets to minimise the shortcomings in your writing designed to persuade – guaranteed.*

Third tip: be specific not generic. People like precision, facts, numbers, statistics. And the more tangible those are, the better. So instead of 'Save up to 40%', it's more effective to say 'Save as much as £65'. The headline gets another tweak:

> *Seven secrets to minimise the shortcomings in your writing designed to persuade – guaranteed.*

Fourth tip: talk about solutions, not problems. People buy solutions, they don't like hearing about problems. So, in an anti-dandruff shampoo ad where they have one big picture, it'll never be a picture of the problem. It'll be a picture of someone with dazzling, flake-free hair.

So the headline gets another couple of tweaks:

> *Seven secrets which can make your writing irresistible – guaranteed.*

Fifth tip: features tell, benefits sell. So, when you get a credit card, 'Fraud protection' is a feature, 'You can feel safe when you shop online' is the benefit.

The changes we've made to our headline have already made it more benefit driven, but we can enhance it further. First, by putting the definite article at the beginning. And secondly by being bolder and saying the benefits 'will' rather than 'can'. We'll also add an important, new benefit: that these secrets are 'easy'. So now we have:

> *The seven easy secrets which will make your writing irresistible – guaranteed.*

Sixth and penultimate tip: make your writing lyrical and lively. Use slightly unusual language, a little alliteration, a metaphor, onomatopoeia or a play on words to make it stand out.

We can make the headline more alliterative by changing 'easy' to 'simple'. And we can make it more distinctive by adding a punchy adjective:

> *The seven simple secrets which will make your writing bloody irresistible – guaranteed.*

Seventh, final tip: ooze credibility. There are lots of ways of sounding more credible – such as quoting someone. Interestingly, it works even if your audience doesn't know the person you're quoting. In fact, it works even if your quote doesn't say who it's from, but just *looks* like a quote. So, let's add speech marks:

> *"The seven simple secrets which will make your writing bloody irresistible – guaranteed."*

And there it is.

From *A few techniques to minimise the shortcomings in writing designed to persuade* to "*The seven simple secrets which will make your writing bloody irresistible – guaranteed*" with a few simple advertising parlour tricks.

Of course, your business writing involves a lot more than just a simple headline.

Your writing tackles complex subjects, is read by a very smart, demanding audience and is competing with thousands of other communications. So, while a bunch of wordsmith tricks can be useful (and there are dozens more throughout this book), we really need to begin by looking at something much more fundamental.

The planning behind your writing.

Sometimes it's tempting to just start bashing away at your keyboard without spending much time thinking about what it is you're trying to achieve. But as they say in the army, 'Proper planning prevents poor performance'.

1

Powerful business writing: Your objective

I've divided the essential planning of powerful business writing into four parts. The *proposition*, the *content*, the *context* and the *concept*.

But before we look at them individually, you and I need to agree on something.

What our *Planning*, *Doing* and *Reviewing* are trying to achieve. Because that 'executive summary' is then your goal which the rest of this book will help you achieve. So, I believe powerful business writing does five things. It:

1. Feels easy and natural for the audience to read and agree with.
2. Tells a story which keeps the audience utterly engaged throughout.
3. Harnesses a style which suits the subject, the audience and the medium.
4. Makes the audience feel glad they read it.
5. Persuades the audience to think, feel or do whatever it is the writer wants them to think, feel or do.

There are also ten things which great business writers *don't* do:

1. Think a complex product/service/idea needs a complex explanation.
2. Write about too many things at once.
3. Start slowly and continue vaguely.

4. Write to impress, rather than to communicate.
5. Write for themselves, rather than their audience.
6. Create ugly work, rather than communications which are as appealing to look at as they are to read.
7. Ignore the advantages and limitations of their medium.
8. Write with too much tone/too little tone.
9. Be purely rational and forget the emotional.
10. Simply inform, rather than persuade.

In fact, I'll summarise powerful business writing in nine words:

> *Powerful business writing tells a simple, engaging, persuasive story.*

That's what your goal should be. And that's what this book is about.

We're going to look at writing in different media, the usefulness of concepts, psychological principles of persuasion, how to develop structure and style, language tricks and a whole host of other advanced writing techniques in different chapters throughout this book.

But, ultimately, they're all about achieving that one sentence. Telling a simple, engaging and persuasive story. However – maybe you don't agree. So let me just briefly tackle each of the four elements of that distillation: *simple, engaging, persuasive* and *story*.

First, *simple*

Now this is a pretty basic (and fundamental) point. But it's one that sometimes seems to be controversial. Because so much business writing appears to be deliberately written to be complex.

It may be that the writer thinks that, because their subject is complicated, their writing needs to be too. And that to be taken seriously, they should use long words and complex sentence construction. That every thought should be expressed in a way that only a clever audience could understand.

They're wrong.

There you go. The effectiveness of a two-word, two-syllable paragraph: 'They're wrong'. The truth is, the more simply and clearly you speak, the better you will be heard.

You've heard of 'legalese'. Speech or writing composed by someone who's been in law so long, they're unable to talk normally. Or the way policemen speak. They don't say 'got out of the car', they say 'exited the vehicle'. They don't say 'house', they say 'premises'. 'Kids' become 'juveniles'. And policeman don't 'walk', they 'proceed'. Everything's that little bit more wordy than it needs to be.

'Business speak' is your equivalent. And it may be that because many senior managers and directors are terrible writers who use this obfuscating style, others imitate them and the vicious circle continues.

So, don't judge yourself by others' awful prose. Always look for the shorter word, the simpler term. Why? Because it's quicker and easier for your audience to absorb. It's more enjoyable to read. It removes the chance of misunderstanding. And because it's easier to read, it'll be read in preference to any other documents which aren't written as clearly.

Plus, it conveys a clarity of thought that, in itself, is very persuasive.

Still not sure? Well, professional copywriters – people who work at ad agencies and who are paid by big brands to write their multi-million-pound advertising campaigns – spend a great deal of time on this. In fact, one of a professional copywriter's most important jobs is often to 'Make the complex, simple'. They spend time finding the pithiest, simplest (yet accurate) way to express something.

Because *that's* clever. To write about a subject in a way that makes people say 'Oh, I'd never thought of it like that before', or 'I've never really understood it until now'.

Essential example

Look at the instruction manual for any Apple product. Before Apple became so successful, other manufacturers used to have vast, verbose instruction manuals in horrendous detail and jargon-infested pseudo-English.

Apple made it simple. Crisp. Brief. They give you the quickest possible route to getting started and enjoying the product you've just bought from them. (Which they then make as simple to use as possible.)

So be like Apple. Find and express the simple truth behind your subject. That's an achievement to be proud of. And while you're at it, cut out the flab so only the meat is left.

As Einstein said, 'You do not really understand something unless you can explain it to your grandmother'. Or, to quote advertising luminary Dave Trott, 'Complicated isn't clever. It just looks clever to stupid people.'

But a note of caution: you're aiming for writing that's simple, not simplistic. Simplistic means you are 'over-simplifying' something, missing out key truths about it. Simple means finding a way to distil the complexities of a subject into something anyone can understand, using everyday language.

So: if you have something to say, say it as clearly and powerfully as you can. A simple, uncluttered style that doesn't stray off the point or fill the page with irrelevant asides is key to that.

Avoid jargon, acronyms, obfuscation, ambiguity and verbosity. Embrace clarity. Aim for writing that is *dense but light*. Dense, because it has a lot of meaning and substance in every sentence. But light, because it's still easy to read, not turgid and taxing to wade through.

It's not easy to achieve. But it's a cornerstone of good business writing. And it'll make your communications stand out from most of what's in your audience's in-tray.

Essential tip

Look at your first sentence or paragraph. What would you lose if you removed it?

Often, the answer is 'not much' and you'd be quicker into the meat of the story. Many people often 'warm up' to their theme as they start writing, but then don't go back and delete their warm-up exercises. So have a look: maybe your third line is the real, powerful opening you need, and you can just cut the two before it.

Secondly, *engaging*

Just writing the *substance*, thoroughly and accurately, can be a lot of work. To write that substance with an engaging *style* is more effort still. So why bother?

Well, most people are exposed to over 4,000 emails, articles, ads and other messages every day. They haven't got time to pay attention to all of them. So they give less attention to messages that interest them less. In other words, no-one's going to read what's dull. You can't bore someone into listening to you or agreeing with you or buying from you.

No, you need to be engaging. And you do that by writing with substance *and* style. In fact, by writing with a style that *serves* the substance. Communications that only have substance (content) are dreary, turgid affairs.

And we've all read stuff that's just substance. Where the writer is so in love with their subject, they don't bother trying to make it interesting and tasty. They just ladle it out in thick, unappetising chunks and they never bother to add seasoning.

On the other hand, you'll see plenty of stuff that's written with very little substance. Some of the business books I've read, for example, have been disgracefully thin on content – just vague, wishy-washy drivel that didn't really tell you anything.

Good communicators find out what's most interesting about their subject, and their writing is filled with great substance. If you haven't got anything interesting to say, go find something. Compose your argument using the best quality ingredients.

Meanwhile, you also see plenty of writing that's just about style. It's particularly common in the work of 'wannabe' writers: people who think they have a gift for writing and want to impress you with their wordplay.

That's a very unhelpful starting point – to write with the aim of impressing your audience with your writing ability, when you should be trying to impress your audience with your content.

So: style and substance. Get the balance right and you'll communicate elegantly and *engagingly*.

To go back to the food analogy, think of your communication as a stew. Too big a portion (substance) and you'll give your audience indigestion. Too little, and you'll leave them hungry and unsatisfied. No seasoning (style), and the meal is bland. Too much seasoning and it's inedible.

As for what 'style' is, well, it means all the writing craftsmanship we're going to cover throughout this book, to give your business communications life and energy.

Essential tip

Using terms like 'passionate about', 'innovative', 'leading', 'world-class', 'consumer focused', 'goal oriented' and so on does not make your writing engaging. It makes you sound like a contender for this year's 'Bullshit Bingo' champion.

Thirdly, *persuasive*

There's a lot more on this topic throughout the book. Psychology in language; how to persuade through writing.

Why is it an important tenet of powerful business writing? Well, if you're not writing to persuade, what are you doing? Just trying to inform? To clarify? To objectively present some facts?

I don't think so.

Of course, you do want to present information clearly, but only as part of a larger goal: to get the reader to agree with you. You're presenting a case for action you want them to take or action you want them to not take. You want them to invest in something, decide something, think something or feel something.

In short: you want something from your reader. And simply 'presenting the case objectively' is leaving it to chance. It's relying on your audience being predisposed to agree with you or being sympathetic to your cause or just weighing up all the facts rationally. Which, believe me, they don't.

No, what you're writing about is too important for such imprecision of purpose. It's not just what you say that can influence your reader, it's the way you say it.

So we're going to say it in the most powerful, influential way possible. Which is why, when we get to the *Doing* section of this book, we'll look at 12 psychological triggers and how to use them in your business writing.

And finally, *story*

Before Wikipedia, knowledge used to be passed on through generations with stories. When you're growing up, learning to read, it's stories you first learn with. And when you see a great public speaker, they'll almost certainly pepper their talk with memorable anecdotes. Stories.

I'm not suggesting that erudite, eclectic business writing follows a *Once upon a time* motif.

But powerful business writing does have an element of storytelling to it. In its structure and narrative. In its personality. In the way you use examples or your own experience and observations to bring your message to life. In the way you make every communication unique, distinctive and memorable.

And in the way your communication builds to an irresistible denouement: getting your reader to do what it is you want them to do.

OK. So we're agreed. Powerful business writing tells a *simple, engaging* and *persuasive story*.

It achieves the five things I listed at the beginning of this section, and avoids the ten things I listed after that.

Now that we know what makes for powerful business writing, we can plan and prepare to achieve it, by looking at the four pillars of *proposition, content, context* and *concept*.

And we'll begin with the most important question of all:

What the hell are you trying to say?

2

Proposition: The thin end of your wedge

Watch a puppy in a new house. It bounds around, tail wagging, tongue lolling, ears flapping up and down as it darts everywhere. Enthusiastically sniffing everything.

Then – woah! It's entranced by your slippers. Your slippers are the most exciting, most important thing in the – wait, what's this? An old newspaper! Yes, that's the most amazing thing I've ever seen! I could play with this forever, I – ooh, a tennis ball!

Business writing can be like that. All over the place. Lacking direction and purpose.

But good communications are written to meet a specific, stated objective. Which sounds obvious – but there's plenty of work that seems written for no reason other than to fill some white space. Or because 'Bill needs a report on his desk by the end of the week'.

You need to know what the aim is. A clear, identified and agreed objective which you stick to like a dried cornflake to a bowl. Before you start writing, make sure you know what this objective is. Have it written somewhere clearly, something like: 'The aim is to persuade our audience to invite us to pitch for their account'.

If you don't know what the objective is, one thing's for sure: you haven't got much hope of achieving it.

Make the objective the very first thing you write, and run it by whoever else is involved first, to make sure you're all in agreement about what it is you're trying to do. You may find people have different views on the objective – in which case you'll save a great deal of time and hassle by getting it agreed before you start writing.

Essential tip

There's a well-known acronym for personal performance objectives – SMART. 'Specific, Measurable, Achievable, Realistic and Time-bound'. They're useful here too: is your objective SMART? If it's not, get one that is.

Now. To turn that blunt objective into the sharp point that will drive home your communication, we'll borrow an idea that's on every marketing and advertising brief in the world.

The proposition.

It's often said that 'What you say is more important than the way you say it'. That's not always true of course. A great orator with mediocre content can often move people more effectively than a poor speaker with great content. Just as a rubbish singer with a lot of enthusiasm can be a more entertaining karaoke star than a half-decent singer with no personality.

In fact, in business – a world of me-too products and services where we're often talking about the same thing to the same person using the same media – the *way* we tell the story may be the *only* point of difference.

But in most circumstances, *what* you say is generally regarded as the biggie. Content is king.

And that content should be built around a proposition. Just ask yourself: if my audience is only going to remember one thing from my communication, just one single thought, what should that one thing be? That's the proposition.

The single-minded, single most important message you want your communication to convey. More colloquially, you could call it 'the elevator pitch'.

Imagine the person you want to persuade steps into a lift at the same time you do. But they're getting out in a couple of floors. So you've got eight seconds to tell them why they should listen to you.

Essential example

Imagine you're the marketing director at one of the big tour operators. One that does Balinese package holidays, which have been popular with honeymooners.

Trouble is, newlyweds are a fairly small audience. So, you've got a declining market.

Your objective is simple: persuade more people to go, not just young couples who've recently tied the knot. People in their forties and fifties (who often have a higher disposable income) would be ideal. They're people who enjoy great beaches too, but they also value high standards of service.

So . . . since it's a great honeymoon destination but you want to entice couples in their forties/fifties, why not position it as a great place to enjoy a second honeymoon?

And your proposition becomes:

Enjoy a second honeymoon even more memorable than your first – with a holiday in beautiful Bali where every day is a special occasion and every guest is treated like royalty.

It's not a headline. It's a blunt form of words that may never appear like that anywhere in what you or anyone else writes. But it is, in effect, what you'd want someone to say the communication was about, if they read it. It's the 'take out' you want them to go away with.

A second, simple way of arriving at the proposition is to consider all the information from a problem/solution perspective. What is the 'problem' your audience has? What is the solution to that problem that you're offering? State your proposition in those terms.

For the Bali piece, we might say the problem for our audience is that holidays have lost their romance. Their magic. Bali, a 'second honeymoon' destination, is the solution to that. And that problem/solution dynamic gives us a clear way to express the proposition.

But like I say, write your proposition bluntly. Without embellishment.

Sometimes people have an urge to write them as if they were a slogan. I saw one the other day: 'Expect the unexpected'. It was about a bank account. Yet the core benefit/solution of the account wasn't that it was 'constantly surprising'. Yes, the audience might be surprised by the benefits, but it doesn't tell them what those benefits *are*, which is much more important.

My subject is much too complicated to summarise in a sentence.

Is that what you're thinking? That you can't write a one-sentence proposition for your communication? Because there's not one important message – there are eight, all of equal importance? Well, remember that your audience may not remember eight things. They may only remember one. So it's better you decide what that one thing is, rather than leave it to chance.

And don't tell me your 'one thing' is too complex to boil down into a sentence.

It's said that an editor of *The Sun* wouldn't let any story run unless the journalist could express that story in 20 words or less. Using vocabulary a nine-year-old could understand. See for yourself: have a look at today's front-page story of *The Sun*. Look at the headline, the subhead and the first sentence.

The headline tells you something. The subhead tells you the same thing again, just using different words and adding a bit of detail. And the first sentence tells you again, adding a little bit more detail yet. And that first sentence summarises the whole story that's then told fully in the rest of the article. So that first sentence is usually 20 words or less.

Essential example

Here's one from 2011 about the tsunami that struck Japan. Headline: APOCALYPSE JAPAN. Subhead: Thousands feared dead in tsunami. First sentence: A MASSIVE earthquake has devastated Japan – sending a 33-ft tsunami smashing into the country's north-eastern coast.

If *The Sun* can summarise any world event in 20 words or less, the story you want to tell can probably be expressed as pithily too. For a business document, I reckon a good proposition can be up to *30* words long (like the Bali one above). But that essence, that pure distillation of your whole communication, is what you're looking for.

As David Selznick, former head of MGM, used to say, 'Write your idea on the back of your business card. If you can't, you haven't got an idea.'

Essential tip

Once you know your proposition, the one thing your audience must remember, write it out with a thick black marker and stick it somewhere prominent.

It's easy to get into the flow of your writing and wander off-topic. Having the proposition staring you in the face as you work helps make sure that doesn't happen.

3

Content: Interrogation & insight

You have a proposition – a clear idea of what you're going to write about.

Interrogation simply means gathering together the content you want to use to bring that proposition to life. All the supporting material that helps you craft that proposition into a compelling communication.

It is, if you like, the logic – the facts, figures, case studies, research findings, experience and expertise that colour-in the outline your proposition describes. Insight – that's the magic. Where, as a result of exploring what you want to say, you come up with a flash of inspiration as to why your audience should be interested. A new way of positioning your subject.

I should just point out, you won't always do proposition, interrogation, insight in this order. In fact, it might be that you've had a great insight – a eureka moment – which is the reason you're writing whatever it is you're writing in the first place. Or it may be that by interrogating your subject, you come up with some interesting nugget that alters the proposition.

Proposition, interrogation and insight are all related – they're all grist to your mill, to help you create a communication that best achieves your objective. But often they'll work in the order suggested here.

Interrogation

This is the marmite bit.

Some people love doing research, some people hate it. Both have their pitfalls. If you just rely on what's in your own head, you're likely to end up with something rather generic. If you simply rely on the same background material everyone else already has, you'll write something that sounds too much like what's gone before. You won't get any stand-out.

So it's important to find out more about your subject. Even if you're an expert on it, there's always more interesting stuff out there.

On the other hand, you can have too much of a good thing. If you do too much research, you'll get lost in minutiae and end up with lots of obscure stuff. Then you'll tie yourself in knots trying to choose what to include and what to leave out. And you'll have left yourself less time to write it.

So, find out what you can *in the time that you have*. You're going to be the person who expresses things in a more interesting way than your colleagues. But also be the person who finds more interesting things to say than your colleagues.

Essential tip

Find out something that your audience won't know. Something interesting, factual, surprising, shocking, illuminating.

If the proposition is 'the elevator pitch', here you're looking for 'the water cooler' moments – some little nugget that you could imagine people sharing with each other as they bump into each other at the office water cooler. Some 'Did you know' titbit.

You'll see plenty of reports, proposals, presentations, reviews and so on that just regurgitate the same old facts, opinions and terms. Find something new to say, and a new way to say it.

Get on the Internet and keep digging until you unearth something more interesting. Something your audience won't know. Collect up the nuggets you find in a little folder (electronic or physical) somewhere. These little gems are going to make your writing sparkle.

There's a second, rather wonderful way of interrogating the subject to come up with new, interesting nuggets about it, of course. And that's to use it/experience it/investigate it yourself.

In the case of the Bali holidays we looked at earlier, you'd go on a holiday to Bali (not too much of a hardship) and you'd get so much rich material. Your holiday journal would capture moments that you'd probably never unearth any other way.

And by later reliving some of those moments in your writing, you'll create a much more authentic, compelling story.

Essential example

Famous ad man David Ogilvy wrote advertisements for Rolls-Royce. So he had a Rolls-Royce, so he could 'interrogate the product'. Which made it much easier for him to find interesting nuggets to bring his ads to life.

We might imagine that the proposition for Rolls-Royce was something like 'The most refined, well-made car in the world'. After driving his own Roller, what nugget did David unearth to use as the headline to bring that proposition to life? 'At 60 miles an hour the loudest noise in this Rolls-Royce comes from the electric clock.'

Or here's a more prosaic example: say you needed a new angle to market the *Banana Guard*. You know this thing? A yellow, plastic, hinged container for putting bananas in.

But let's 'interrogate the product' – play with it, examine it and imagine what all the benefits could be.

The main benefit it's promoted for is protecting your banana from being bashed around in your bag. I bought one not to protect the banana, but my bag – after a forgotten banana dissolved in a previous bag, ruining it.

So that's a second, complementary benefit for a start. Here's a longer list you might come up with:

1. protects your banana from damage
2. protects your bag or clothes from banana stains
3. works on virtually any size/shape banana (fits 95 per cent)
4. holds the skin afterwards if you've nowhere to put it

5. holes keep fruit fresh (ventilation prevents premature ripening)
6. bright so it's easy to find
7. looks like a piece of banana art and looks good in fruit bowl
8. acts as a visual reminder to get one of your five a day
9. can hang it up by one of the holes so it's out the way (eg, off your desk)
10. lasts a lifetime (approx. 6000 bananas), and at £6 that's .1p per banana!
11. stops other fruit from over-ripening by being close to banana
12. also works with some cucumbers, celery sticks, carrot ...
13. subtle branding keeps it classy (but reminds you of name in case people ask)
14. unlike most kitchen products, no sticker that's impossible to get off properly
15. can be used stuck in a pocket as a pretend gun
16. light: only two ounces makes it easy to carry around
17. compact: takes up little more space than a banana, so fits neatly in a handbag or suitcase
18. makes you look wise and healthy
19. encourages people to eat more bananas because they're no longer worried about carrying them: studies show people with a Banana Guard eat 18 per cent more bananas than they did before
20. dishwasher-proof so easy to clean
21. comes in different colours: how about a blue banana! or a green one!
22. has some friends: the *Pear Guard* and the *Froot Case* so you can collect the whole set
23. made from FDA-approved recyclable plastic so they're environmentally friendly.

Et voilà: you have a longer list of benefits than the makers of the product have ever promoted, and more interesting things to say.

Benefit 19 – that people with a Banana Guard eat 18 per cent more bananas – is made up. The point is, by spending some of your planning time interrogating your subject, you might come up with an angle that you think would be really powerful. You just need to then get the data to see if you can back the idea up.

Insight

'Eureka', said Archimedes. Upon spilling bathwater and Radox all over his travertine.

That moment – when you turn a fact, idea or bit of knowledge into something you can use – that's an insight.

To put it another way, if your proposition is the *what* you're talking about, the insight may be the *why* you're talking about it.

And when something occurs to you – something new – as a result of your subject interrogation, that's your insight. Your eureka moment.

Many strong insights are *audience* insights – something you've discovered about the audience that helps show why your subject is exactly what they should be interested in.

I say 'discovered' – some insights you can find out by reading the research or by information gathering, or by watching a focus group or holding a co-creation session. But many of the best insights are intuitive.

In Schiphol airport in Holland, for instance, they wanted to get men to improve their aim at the urinals. They couldn't really run a focus group, asking men about their peeing habits. Nor could they stand in the men's toilets with a clipboard, observing what went on.

They had to use intuition into what men are like to come up with an insight into what would work. What they did was engrave a fly on the urinal. Near the plughole. And men suddenly had something to aim at. Misfires went down by 80 per cent. Which saved on cleaning costs and made the toilets nicer.

Because someone had the insight that if men saw a fly they'd aim for it. Try putting yourself in your audience's shoes and considering how they think, feel and act. Use what you do know about them to take that next step: something you don't know for sure, but which your intuition tells you might well be the case.

And see if that gives you a great angle for positioning your subject.

Essential example

When advertising creative Alec Brownstein wanted a job at a big agency in New York, his intuited insight was that creative directors in New York were probably very vain. He thought they were probably so vain they googled their own names on a regular basis.

So he used Google Adwords. And when any one of New York's five most prominent creative directors googled their own names, his one-line copy ad (saying it was fun to google your own name and also fun to hire him), appeared above the search results. Four of them ended up interviewing him. Two of them offered him a job. He took one of them.

His ad had cost $6.

Let's go back to the Banana Guard: we have the list of product benefits from our interrogation session earlier. Benefit 4 is: gives you somewhere to put the skin afterwards.

Why would that be useful? Only when you're somewhere where there isn't a bin handy. Such as in the car.

So there's a bit of intuitive insight: we can use Benefit 4 as a reason to appeal to people who travel in cars (ie, an awful lot of people). Have a Banana Guard handy and you can take fruit in the car and have somewhere to put the skin afterwards, without it rotting in the ashtray or without you naughtily throwing it out the window.

In fact, taking it further, our insight might be to recommend the company sells the Banana Guard at petrol stations/motorway services, with Benefit 4 clearly flagged on the tag.

And that's what insight is. A great deal of business writing ignores it, just sticking to dry facts. But a little insight can make all the difference. It's the mental leap which turns information into an idea. Or a report into a discovery. Or a presentation into a standing ovation. Or a proposal into a promotion.

Which is why it's well worth spending some of your planning time trying to find an insight into your audience or your subject that will lift your business writing.

Essential example

Invention of the 'Post-it' Note. It was a guy called Spencer Silver at 3M who invented the low-adhesive glue (by mistake). He probably wrote lots of reports about it. But he couldn't find a use for it. It was Art Fry, attending one of Silver's seminars, who came up with a use for it.

He had the same facts Silver did. The same overall objective: to develop new products for 3M to market. But it was Fry who had the insight into how it could be used.

And forever more, it was Art Fry who was known as the inventor of the Post-it Note.

Context: Audience & medium

Audience

Our planning is going beautifully. We know what we want to say. Why we want to say it. And what we hope we'll achieve by saying it.

Now, who are we going to say it to?

Allegedly, when Coca-Cola first entered China, it was translated as *Ke-kou-ke-la*. Which means 'Bite the wax tadpole'. Eventually they found a close phonetic equivalent, *Ko-kou-ko-le*, which means 'Happiness in the mouth'.

Pepsi's luck was no better. Their slogan 'Come Alive with the Pepsi Generation' became 'Pepsi brings your ancestors back from the grave'.

And in Africa, where there's a large proportion of people who can't read, it's common to put a picture of the contents of a tin on the label, so someone illiterate knows what's inside. Gerber didn't know that, so their baby food … had a picture of a baby on the side.

Which goes to show: it's important to know who you're talking to. What they'll easily understand. What's appropriate for them. And what will resonate with them.

You might be writing to one person you know personally. Or one person you don't know. Or a dozen people who all have very different personalities. Or thousands of people, who you can't possibly know.

Which of those audiences you're writing to should affect the way you plan your communication.

Writing to someone you know

I'm sure you've heard the term 'mirroring' from body language. It's the idea that if two people are adopting the same body language, they're in sync with each other. And that you can therefore mirror someone's body language to make them subconsciously feel that you're on the same wavelength as them. It creates a stronger connection.

You can do the same in business writing. How? By using a little of the favourite phrases and constructions of the person you're writing to. If they use the phrase 'knock it out of the park', then perhaps you could too.

Plan to use these phrases judiciously and you will literally be 'talking their language'. Which is a good thing. If they notice it, they may well be flattered. If they don't notice it, they'll still be subtly influenced.

Only do this sparingly, though – you don't want your writing to seem like a parody of theirs. Just save it for your key moments. It may make your ideas seem more like *their* ideas, which means they're more likely to agree with them.

When you're writing to someone you know, you'll have a good idea of their prior knowledge too – what you won't have to go over in detail, what technical phrases they'll be comfortable with.

Because, although every guide on good writing will tell you to avoid jargon, it's not a good idea to avoid specific terms that are common in your industry and which you know your audience will be *au fait* with.

But, do follow the convention for acronyms – spell it out first with the acronym in brackets after for the first use. For example, Return On Investment (ROI). After that one use, you can just refer to ROI.

I've seen, on many occasions, documents being browsed by a room of senior managers where half of them don't know what one of the acronyms being used is. That alienates them (undesirable) and means they don't take in your communication properly (very undesirable). So don't assume – spell it out once the first time you use any abbreviation.

Essential tip

Use the person's name. If you're writing to one person, for them alone, not for them to pass on, then that kind of personalisation can be very powerful. Don't throw their name around like confetti at a wedding, but use it every so often to maintain that personal, conversational link.

Writing to someone you don't personally know

I'm going to (kind of) lump everything else into this category. Because most business writing – most writing full stop – is to someone you don't personally know.

I don't personally know you, for instance. But I do know something about you. Simply because I know the 'target audience' I'm writing this book for. Business high fliers. Senior managers, directors, entrepreneurs, executives, business owners, captains of industry.

Basically, bright, motivated, business-savvy people who know that being able to write powerfully and persuasively is a tremendous competitive advantage. People who want greater success, and who are prepared to work hard to achieve it. People who want to learn more, even though they already know a great deal.

People, in other words, who would want to read this book. So I shape the writing to suit the audience. The tone, the references, the content, the examples, the language – all chosen to suit you.

Which probably sounds incredibly obvious. Yet in business, most people write everything they do *in exactly the same way*, regardless of who they're writing to. But your personality when you're not writing business communications isn't set in stone. You adjust your personality and behaviour according to who you're talking to 'in real life'.

While you're always 'you' with your own particular nature – your 'brand', let's say – you probably talk to your mother differently to how you talk to a friend. Which is different again to how you talk to your partner, which is different again to how you talk to your boss.

In the same way, your copy will be more successful if it's tailored to suit the person you're writing it for.

So, a) find out as much as possible about who you're writing to, b) put yourself in their shoes and c) write for them, in a style they will appreciate, with substance that will interest them.

Find out as much as you can about who you're writing to

'Let the dog see the rabbit'. In other words, find out as much as you can about your audience. Because, when you can see them in your mind's eye, you'll be better able to write in a way you know will connect with them.

You may have seen marketing descriptions of an audience that look like this: 'ABC1 empty nesters who read *The Guardian* and like cookery programmes on TV'.

Which means they're in socioeconomic group A or B or C1 (ie they're middle-class, white-collar workers), their children have left home, they read a left-leaning broadsheet newspaper and they like Jamie Oliver. Now do you know who you're writing to?

Probably not. That kind of marketing jargon hardly puts a real person in your mind's eye. What can be more useful is a 'pen portrait'. This describes the target audience a bit more fully with an invented example: a name, a picture and a description of what the target audience is actually like. A few paragraphs done in a 'day in the life of' style, for instance.

Find out what you can about the audience for your writing, and create your own pen portrait for them in your planning. It will help you 'put a bit of flesh on the bone'.

Essential example

You've probably heard of Neurolinguistic Programming (NLP). There's an idea in that which says people often have a 'visual', 'aural' or 'kinesthetic' (touch) processing preference.

So someone visual is more likely to say, 'I *see* what you mean' or '*Look*, what I mean is . . .'. Someone aural is more likely to say, 'I *hear* what you're saying' or '*Sounds* good to me'. Whereas someone kinesthetic would say, 'I really *feel* that . . .' or 'I *grasp* what you're saying.'

So if you're writing to one person you don't know, maybe you could find out – from their writing or speech – which of VAK (visual, aural, kinesthetic) they prefer. Then use that form in your writing and perhaps they'll find it easier to 'process' your communication – which could make them (subconsciously) more likely to respond to it.

Put yourself in their shoes

To add more colour to your pen portrait, think about the person you personally know who fits your target audience, and imagine you're them. Think about their likes and dislikes. Their views. The way they talk. What gets them nodding along, what interests them and what winds them up.

Write for them

Then write as if you're talking to them. More than that: as if you're trying to persuade them.

Essential tip

Always, always write as if you're talking to an individual, rather than a group, even when you are talking to a group. So don't say 'We hope everyone will . . .', say 'I hope you will . . .'. Don't say 'People like you . . .', say 'You . . .' Don't say 'People at this company have often believed . . .', say 'You may have believed . . .' Talk to someone as a singular entity, not part of an amorphous mass.

Something else to write out for yourself, to go with your proposition, can be an 'audience journey'. Simply write two things. First, what you believe your audience *currently* thinks and feels about the subject you're writing about.

And secondly, what do you *want* them to think/feel? Now you can plan what you need to write to move them from the first position to the second.

My favourite example of understanding the audience (and the brand) comes from clothing catalogue company Boden (whose founder, Johnny Boden, apparently writes much of the copy, with a famously idiosyncratic style).

One year they sent out their Mini Boden catalogue and it featured a kid's T-shirt with a picture of a toy revolver on it. They got a lot of complaints. So allegedly Mr Boden sent an email to their customers. Apologising at length for the T-shirt. Saying it was a mistake. Remarking that, 'We feel stupid. Especially me.'

And what do you know? The next catalogue achieved record sales. The audience felt Boden understood them. Shared the same values. And, most of all, listened to them.

In fact, I spoke to someone who used to work for Boden who said that after that they considered whether or not they should find *more* reasons to apologise to their customers, since it seemed to be so good for sales.

One final point on writing for an audience. Sometimes you might have to create a communication that's perfectly aimed at two (or more) distinct audiences. Which is tricky.

Essential tip

The only practical answer is to write your copy to whatever aspects the audiences have in common, particularly tonally. If you put what you know about each audience in a Venn diagram, you're writing to the middle bit where they all overlap. Although, as you can imagine, writing to this more homogenous group isn't ideal, and can knock the edges off your copy.

In summary, be like a good comedian: know your audience. I promise, it will make your punchlines much more effective.

Medium

Great writing is great writing, whatever the medium. But writing does work differently according to the environment it appears in. Often for simple physical reasons – such as the amount of space they afford, the layouts they suit, the way audiences consume those media, or the things you're able to do in one medium that you can't do in another.

Let's look at the basics of six different media that your business writing might commonly appear in.

Some of them – such as reports and email – you're sure to create. Others, such as business-to-business or business-to-consumer direct mail and business advertising, depend on what your role is or what you want to achieve. However, a quick tour of these marketing/advertising media is still useful to get another perspective on how to craft a compelling communication.

I would also just mention that when you're writing a long document, there are lots of template aides out there. Modern versions of Microsoft Word, for instance, give you a wide range of templates to choose from.

You don't need to be a computer expert to use them, but they helpfully create layout grids for you so you can have the most important information or the most interesting quote alongside the main body of text, for instance. Or so you can insert a picture easily. I'm using Pages at the moment, and that has similar templates I can download from the App Store.

Try a couple of different templates and find one you like. It will make your pieces stand out – and make them more enjoyable to read too.

1 Proposal/report

It may be that most of your important business writing falls into this broad category; documents of several hundred or several thousand words written in a word-processing document and emailed on or printed out. So we'll do the equivalent of a SWOT analysis (strengths, weaknesses, opportunities, threats) on the proposal/report format.

First of all, there are some basics of layout. If your document is going to be A4, then generally your main text should be around 11-point size. Of course, different fonts work at different sizes (sans serif font Gill Sans looks much smaller than Avant Garde, for instance) but 11 or 12 is a good guide size for the font. Your line length (the technical term is 'measure') should be around 14 to 18 words.

Bear this stuff in mind – it may sound trivial, but if your line length is too long or too short, your communication is harder to read. While we're on layout, here are four other basic tips:

1. Vary the paragraph lengths (with a line space between each) but don't have any paragraph longer than five lines. Don't use bold *and* italic *and* underlining *and* capitals – stick to a couple and use them sparingly and consistently (like the way bold and italic is used throughout this book, for instance).
2. Use a consistent range of sizes: main text may be in 11 or 12 point, section headings in 14 point and headlines in 18 point.
3. Keep decent margins around the page so it looks light and easy to read.
4. Use a consistent scheme for headings: have main headings (all in, say, 18 point), subheads (all in, say, 14 point) and then indented bullets for lists (or numbered lists, like here). Being visually consistent just makes the document feel more professional.

Remember that the purpose of layout is to achieve one thing: to make your writing easier to read, so the reader can concentrate on what you're trying to say. Don't do anything that makes that harder (like putting text over an image) and do make any adjustments that will make it look easier and more elegant (such as bringing the page indents in to reduce the amount of text per page).

Essential tip

We're going to look at headlines in the *Doing* section of this book, but I just want to make the point early that even on a proposal, report, thought piece, strategy document or whatever, your headline should be interesting, engaging and relevant. Just because many business documents have headlines duller than celery doesn't mean yours have to. In fact, some of the best writing examples have great headlines.

I saw an in-depth, scientific essay into the effect the Internet has on the way we think. It was called 'Is Google Making Us Stupid?'. A headline that's short, simple, provocative and intriguing. Or how about this from a major French bank – a report about Chinese growth and an increase in commodity prices called 'The Dragon Which Played With Fire'. It's so exciting it sounds like a Stieg Larsson novel.

As well as headings, subheads and bullets, consider 'call outs'. These are sidebars of text, or indented paragraphs or separate blocks of writing that may be in a larger font. Like in a magazine article, where a key part of the main text is repeated, larger, in its own area of the page.

Subheads, bullet points, indents, call-out boxes, sidebars – they all help break up your writing into bitesize pieces, creating *a skip and dip* opportunity for your audience.

That way, even if they skip some parts, the subheads and bullets will leap out at them. Which means you should plan to put some of your most important 'take-out' messages in those *skip and dip* sections.

Next, consider images.

I'm sure you've heard the research which suggests good-looking people get promoted faster in business. (Taller people too, apparently.) Well, good-looking reports, proposals and so on get preferential treatment too. And selective use of a few images is an easy way to give your document a facelift.

Done properly, anyway.

Back in the 1980s and 1990s, people discovered the 'clip art' that Microsoft included with Office. Suddenly jumble sale posters, window cleaner's leaflets and village fete banners were anointed with these little cartoon drawings. But sadly, so were many business documents.

And they looked awful.

Happily, the world's moved on. Now it's easy to add photographs to your business reports – and it can make a real difference. I don't mean those shots of cheesy, fake American families 'goofing around'. Or the ones of models with perfect teeth pretending to be businessmen and women, pointing at computer screens. Run from those like an ant from a kid with a magnifying glass.

But good photography can make a real difference. I'm not sure of the 'a picture is worth a thousand words' adage, but we do live in an increasingly visual world. And as I say, using a few images in a proposal can make it look more appealing, create more of a feeling (which, as we'll look at later, is important) and make your report stand out from everyone else's.

People on Flickr will often let you use their photography for free, if you ask them – or in exchange for a nominal sum. Royalty-free stock libraries like Shutterstock let you download images really cheaply for a monthly subscription.

The planning stage of your writing is a good opportunity to source a few images. You don't need many – perhaps one large one to begin, then one smaller image for every 200–300 words. You may use images which literally illustrate your point (ie, if you're talking about a new piece of technology, showing that piece of technology). Or you may use images that create a feeling and mood that you want to convey.

So in a report about an opportunity in London, for example, you might use images which are all of London: red buses, the Millennium Eye, the Thames and so on.

But avoid images which are literal. Where you're saying, 'We've tied ourselves in knots over this one' and you show an image of a businessman in a suit literally tied up with a big rope. Or 'We need to expand urgently' and you show a businessman all bent up in a small space.

That kind of imagery is only going to undermine the credibility of your writing.

As with everything else here, go for consistency in your images too. If you use imagery of people, make *every* picture in the report have people in it (unless you want to differentiate between your 'feeling' imagery and 'literal' imagery, in which case the former could always feature people, and the latter never show them).

2 Presentation

If you give presentations you may use presentation software. Like PowerPoint or Keynote or Prezi.

And you may have handouts in the presentation. And/or a 'leave behind' for people to take away after the presentation. Now these are three different communications: the presentation, the handout and the leave behind. They should all be different, since they're doing different jobs. But the most common mistake people make when using something like PowerPoint is *they design their slides to work as a leave behind.*

By which I mean they put the entire content of the presentation on screen, in their slides. And they then spend the entire presentation staring at the screen, reading aloud what's there for all to see.

That's not a presentation. Not a good one, anyway. In fact, it's a trick people use to *avoid* presenting. Because if everything's on screen, you don't need to learn your material. And if you're reading everything off a screen, you don't have to look at your audience.

Great for you. Terrible for your audience. Diabolical for your presentation.

Essential tip

Here's a simple (though not infallible) guide. Look at your presentation slide, and ask 'Does it make complete sense, without any explanation from me?'. If the answer's 'Yes', then you've probably got too much info on the slide. Because the slide should illustrate and add to what you're saying – not *replace* what you're saying.

Generally, presentation slideshows work best when a slide has:

- just one killer fact on it, or
- just one illustrative image on it, or
- just one single line of text on it, or
- just one topic heading, so the audience knows where you are.

You can have a little more – three facts as bullets or a headline and an image.

But generally, plan to keep the writing on a slideshow very, very short – think of it like a billboard. Which means less than 20 words to a slide.

Essential example

I've lost count of the number of times someone's stuck up a chart on their slideshow and said, 'Now you won't be able to read this, but . . .'. Wait a second – you're showing me a slide you *know* is useless? Bloody hell!

If people don't need to be able to read the chart in the presentation, don't show it. If they do, give them a handout. Then use the slideshow to show the take-out learnings from the chart. Three interesting things we can learn from it? Then put one on each of three slides.

To avoid duplicating work, you could write your whole presentation and print it out. That's the leave behind, as well as your notes. Then print out the detailed graphs, charts, etc. that you want people to be able to examine in the presentation. They're your handouts.

Then go back and take out the absolute essence of your presentation, and simplify it into single lines or three bullet points or an image. That's what you show onscreen during your presentation.

Personally I don't think PowerPoint (or any of the others) look very elegant or professional printed out. Better to write the leave behind in a word-processing program and then copy the distilled text into PowerPoint for the slideshow.

It's much more work. But it works much more.

Essential example

I was in a board meeting not long ago when one of the directors wanted about £50k to buy a licence for something or other. He presented his case using PowerPoint. But not onscreen – his slides were bound into the board report. Four slides to a page. In black and white. It was incomprehensible – and he didn't get the £50k.

3 Email

You may be using email as a way of sending someone a longer document as an attachment. Or you may be using a designed, HTML email that's going out to thousands of people. Or thirdly, the most common use of email in business, when you use it to rattle off a quick memo.

Whichever type of email you're planning, there are a few commonalities between them. Here are four:

1. *The subject line.* If you're not in your recipient's address book, you might get spammed. Especially if you use words in the subject line which spam filters don't like. These vary according to the Internet service provider, but *free, win, cash, sex* – all of those struggle. As do headlines in capital letters (I remember a digital planner telling me about working for 'CASH FOR GOLD', whose name unfortunately suffered from being in capital letters *and* for two of the three words in its name being spam).

   Using the person's name in a subject line can work well – and you certainly want your subject line to be clearly *about the subject*, not trying to be clever. The clue's in the name really: 'subject line'.

2. *The fold.* This term isn't literally true, there aren't any folds on a screen, obviously. But it refers to the cut-off point of your email if someone's viewing it in a preview pane or viewing it full screen but the email is longer than the length of one screen.

   Because, while the statistics vary, they all agree on two points:

   - Many people *only* view emails in their preview pane (which means they're often only seeing the first 25 per cent of what you've written);
   - At least 50 per cent of people don't bother scrolling an email (so they'll only ever see one screen's worth).

   All of which makes it vital that you get to the point quickly. It doesn't mean that you have to say everything you've got to say in only three sentences – it means you've got to put the most important information first.

   So, for instance, instead of those internal emails that present the background, the evidence, the argument … and build up to a recommendation in the last sentence, put the recommendation *first*. Then use the rest of the email to explain *why* it's your recommendation.

To help, consider starting your email with a sentence along the lines of 'Why we need to … ' or 'Three ways we can … ' or 'How to guarantee that … ' or 'What X means to you'. A how/when/why/what which cuts to the chase.

Also consider negative iterations – 'Why we mustn't … ' or 'Don't ever … ' or 'Why I say "no" to … ' or 'When to not … ' or 'No more … '. They're often very powerful (just as negative emotions can often be more powerful than positive ones).

3 *The personal touch*. Email, in theory, is a direct medium. One email address going to another email address. So you know who you're writing to, and they know who's writing to them. So use their name, make references to shared experiences, views or values, and make it sound like you wrote this one email specifically with them in mind.

Essential tip

Remember to put your name at the end, even if your email already has a 'digital signature'. It just feels more genuine.

4 *The response*. If you're planning an HTML email that's being designed by someone who knows email best practice, then they'll know to put a number of links in the email – underlined, clickable text, buttons, links to social media and so on. And they'll be able to track open rates and click-thru rates.

But if you're writing an 'ordinary' email – just text, from one person to another, how do you know if the other person has read it? Whether or not they're planning to reply? If so, when?

The simple answer is to ask them, in your email. Not in a pushy way that's likely to irritate your audience, but in a polite, clear way that lets them know what action they can take (and when by) if they've been persuaded by your message. Which, by the time you've mastered everything in this book, they will be.

Because good salesmen always *ask for the sale*. They don't just present the facts and leave it to you – at some point they ask you, 'Would you like to buy this?'. Find a way to do something similar in your emails. People are bombarded with them every day, and you need to know if yours has got through.

4 Other digital

'Digital media' covers an ever-growing array of formats. But it can be helpfully divided into three types: *bought*, *earned* and *owned*.

Bought means anything you pay for – like ads on other people's websites or searches or sponsored links. Earned means anything that doesn't cost you media space – social media when people are talking about you on Facebook or Twitter or in the comments section of the BBC website, for example. Owned, unsurprisingly, means media you already own – like your own website.

Each type has its own strengths, but one thing that's very useful is to encourage them to interact. Write a blog on your owned media that you can link to (and get talked about) in earned media, for instance.

Social media may be where the action is (as well as the most difficult to get right). But, infuriatingly, it's also the area that can be least directly affected by your own writing. Since, by its 'social' nature, it's more about what everyone else is writing, not you.

The other thing, of course, is that the 'digital space' changes at a rapid pace. Giants of social media come and go (remember Friends Reunited? Bebo?). New ways to reach your audience appear as bright as butterflies, but often turn out to have as short a lifespan.

There are lots of good resources online about writing online. As well as plenty of guides to the different sub-categories within 'online' and what all the terms mean.

Of course, your business writing might not need a great deal of digital involvement. But, just in case, here are five general tips I can pass on from my time of 'doing digital':

1 'The digital space' has quickly become one of the most interactive, involving and (in a virtual way) 'social' mediums of all. People are comfortable with providing a lot more personal information about themselves than they ever have in any other medium at any time in history. As a consequence, they expect you to know who they are, and tailor your content appropriately.
2 Technical innovation sometimes outpaces need. Sometimes it gives you options for fancy-pants effects that don't really add to the story. Be wary not to do things just because you can, rather than because it helps your communication.
3 Write briefly. Shorter words. Shorter sentences. Shorter paragraphs. Snippier snippets. What doesn't look like much on a page looks a lot longer online, an area where people

are click-happy and have very short attention spans. Ernest Hemingway would have been a good online writer: sparse, meaty writing.

4. Don't ignore the technical considerations: consult a cyber-geek (who, in my experience, are always very willing to share their knowledge). For example, search-engine optimisation (SEO) is a whole discipline in its own right, which you may not be interested in learning.
5. So instead, check with someone who knows, for instance, that the key words should be in the headline. That the nearer the front of the headline, the better. And that the title might be best under 72 characters, so it appears in full in search results.

   Ask people to get involved, and they often will (partly, perhaps, because digital media make it so easy to). Say 'please retweet' and you'll get many times more retweets than if you rely on just being relevant and interesting. Asking people on Facebook to get involved in a flashmob or some other event can reap dividends. Ask people to link, ask people to comment.

All of this interactivity is an important part of what online is about, so make it an integral part of your writing.

Essential example

A business getting it right. Greenpeace asked people to forward an email message to Nestlé, asking them to stop using palm oil sourced from environmentally unfriendly suppliers. It worked. Then they asked people to forward a message to HSBC, asking them to stop funding those same suppliers. That worked too.

Essential example

A business getting it wrong. The moderator of the Nestlé Facebook page got it wrong big time. When people altered the Nestlé logo as their avatar on the fan page, the moderator told them their messages would be deleted, along with sarcastic and condescending comments and a lofty 'we make the rules' mantra. It spiralled horribly, with a tremendous backlash against Nestlé's Big Brother approach. A PR nightmare, they eventually had to apologise ... and allow a lot of negative commentary and threats of boycotting to build up, all as a result of their misunderstanding of 'the power of the crowd' that exists online.

Search online with 'Gillian McKeith/Ben Goldacre/Twitter' for another example of getting it wrong, in this case when McKeith (allegedly) badly misunderstood the medium she was using.

5 Mail

Direct mail has a number of things going for it:

- You have a number of elements, so you can tell the story in layers and give different roles to different items within the pack.
- You usually have a letter, often around 700 words, giving you a longer opportunity to persuade the audience.
- Because you're writing to a named individual, you're likely to have some useful insights about them, so your copy can be more tailored.

Essential example

An agency I was at tried writing 'male' and 'female' versions for a client. The male version was more to-the-point, more focused on factual detail, using bullet points to list things. The female version was more storytelling and more emotive, less focused on the nitty gritty.

The male version did better to men than the female one to men; the female version did better to women than the male one to women. And the difference was statistically significant.

The fourth great joy of direct mail is that there's a lot you can do with the medium.

I've done packs that have used 'seed paper' that you can plant afterwards and watch grow into flowers. Packs with holes all the way through them. Packs impregnated with the smell of cut grass. Packs where the envelope was made of 'Tyvek' (untearable paper). Packs where something flew out when you opened it.

I've done mailings that played audio when you opened them. Ones with all sorts of different die-cut shapes or bits of embossing or scratch-off areas. Big packs, little packs, thin packs, fat packs. Digitally printed packs (where you can personalise every image, if you want). A one-piece mailer stuck to a piece of astro turf.

Essential tip

In a mail pack there's almost always something more interesting you can do with a piece of paper than make it a leaflet. Make 'What can we do that's more interesting than a leaflet?' your starting point.

Spend time on how you can use the medium in a way that's interesting, but relevant to your message.

The fifth great advantage of direct mail is that everything you do is measurable. You can run one idea against another and see which gets the most response. You can test a shocking idea against a comforting one. Long letter versus short letter. One offer compared with another.

Essential tip

Think of the outer envelope as a press ad space. If you do a 'reverse flap' outer (one where the window for the address to show through and the envelope's flap are both on the same side) then on the other side you've got a nice space, whether portrait or landscape, to create intrigue or provocation or make a promise.

One thing to ensure with direct mail is that your concept tracks through the whole pack.

I've seen too many mailings where the idea is entirely on the outer envelope; you open it and it's just by-the numbers junk mail with scant regard for the concept it began with. Think about how every element can contribute to the concept, which stems from the proposition. Combined with a well-written, personal letter, it can make for a very effective medium.

Essential tip

Just as you may have a hierarchy of information, have a hierarchy of size in the mailing too. Put the most important stuff on the biggest element (usually the letter). Put the least important information on the smallest element in the pack.

6 Press

Double-page spread, full page, half page, quarter page, 20 double, feature-link, advertorial ... these are just some of the many types of press ad.

Height (in centimetres) is first, width (by number of columns) is second. So '20 double' means an ad 20 centimetres high, two columns across. The columns are the columns that the publication divides its pages into (creating a grid for its layouts and columns of copy for its articles). Look at *The Times*, for instance – every page is divided into five columns of type.

A feature-link ad in regional press means you supply info for a journalist to write an article (on a topic that's relevant to your ad), and in a moment of great serendipity, your ad appears right alongside the feature.

An advertorial is an advert designed to look like an article in the paper, usually given away by having ADVERTISEMENT FEATURE written across the top.

What you do in the space you have depends entirely on your brand, audience, subject and your imagination. If you've an image, a headline and maybe even a reply coupon, but only a half-page ad, then there's not going to be much room for more than 20 or 30 words.

On the other hand, there have been some fantastic full-page ads that have been 800 words long. Some longer than that. It's worth remembering that, if someone's reading a newspaper or magazine, then they're comfortable reading a longer ad. In fact, they bought the publication because they *like* reading.

So as long as your writing is utterly captivating for that audience, there's no reason to imagine that tall towers of words will put them off.

Essential tip

If you're doing a classic headline + image press ad, hold dear to the 'one straight, one twisted' maxim. So don't make both headline and image utterly straightforward, or you'll have a dull ad. And don't make both the image and the line clever, or you'll have a confusing ad.

Instead, if you have a clever headline, have a straightforward, clear image. Or if you're using a 'witty' image, have a straight, clear headline.

For example: one of the classic VW Beetle ads had the headline 'Lemon'. That intriguing headline is the 'twisted' part, so the image is 'straight'; simply a picture of the Beetle.

Conversely, a great, award-winning ad for a VW Polo many years later has a witty image instead. It's a wedding photograph where the bride and groom are out of focus, because the photographer has focused on the bus in the distance behind them, which has the line 'Polo L £8,240' written on the side.

It's a clever concept – the idea that the price is so astonishing, the photographer is focusing on that rather than the bride and groom. So this time, because it's the image that's 'twisted', the line is straight: it simply says *Surprisingly ordinary prices*.

You don't have to have an image and headline to make it a good press ad of course. One of my favourite series of press ads was for Tesco. One of the ads, for instance, just had a picture of an apple with no headline. The copy read: 'Granny Smiths. What's the difference between ours and our competitors'? Not much really. They're the same quality as Waitrose. And the same price as Asda.'

5

Concept: Bringing your story to life

All your *Content* and *Context* planning has gathered together your eye of newt, phlegm of bat, dried mandrake root and the sigh from a melancholic mermaid.

But to create something truly magical, you should consider a concept to bring the ingredients to life.

And to help with your hubble, bubble, toil and trouble, I commend two very useful allies to you: SOPHIE and BOB. Just before we bring in SOPHIE, let me make the obvious point that not *everything* you write needs a concept. If you're writing an email to a friend, for instance, you'd be unlikely to give it a concept. Unless you were the sort of person whose dinner also came with a concept.

A concept is an idea borrowed from advertising. TV ads, billboards, viral campaigns – they all have concepts. And they're developed in ad agencies by young creative teams with modern haircuts and ridiculous trousers.

So you may wonder what relevance a concept has in business writing. Well, basically it's a theme that lifts your presentation, proposal or report from humdrum to humdinger. In business writing you'll use a concept a lot more subtly than an ad campaign would – but the principle's the same.

An ad campaign has a concept because they want to find a new way to stand out from the crowd. To connect with their audience. To add an intangible value to their communication in addition to the content of their communication, in order to be appreciated, chosen, bought. And isn't that what you want from your business writing too?

SOPHIE

OK. SOPHIE is an acronym to sense-check your ideas by. It stands for *Simple, Original, Powerful, Honed, Intelligent* and *Emotive*.

They're all pretty self-explanatory, but I'll cover them briefly, paired up into three couplets.

Your concept should be Simple and Original

This is the toughest pairing of all: coming up with an idea that's both simple and original.

Coming up with a simple idea is relatively easy. Trouble is, lots of the best simple ideas have already been done. The simple way to express the emotional benefit of insurance, for instance, is *peace of mind*. As a result, that hoary old phrase has been used on three-quarters of the insurance communications out there for decades.

Coming up with an original idea is also relatively easy, if you just go mental. Chances are no-one's ever made sprout, toothpaste and wine gum casserole before. It would be original. But probably not very good.

No, finding an idea that's both simple *and* original (or at the very least, *fresh*, if original is too lofty an ambition) requires you to sit quietly and scratch your brain with a sharpened pencil.

And if you're brave enough, show people the idea. Do they get it straight away? Or do you have to keep explaining it to them, before they eventually say 'Oh … I see' (in which case it's not simple enough). Do they say, 'Oh, I loved it when Sony did that' (in which case it's not original enough).

Essential example

The tagline for a new Audi sports car: 'Mirror, Signal, Outmanoeuvre'. Three words. Only three letters of which are different from a well-known driving phrase. Really simple. Yet I've never seen it used as the lead concept for a car before.

Your concept should be Powerful and Honed

Powerful and honed just mean making every aspect of the concept *even more so.*

Go back to the first two: simple and original. How could you make your idea simpler? What could you lose from it to make the idea clearer, more immediate? How could you make it more original? What twist or frisson of newness could you add to lift it above the 'seen it before' category? How could you amplify the idea, turn up the volume and make it more attention-grabbing, more engaging, more persuasive?

This is not a time for lily-livered whisperers, humbly suggesting their wares. This is a time for potent, irresistible voices, for mighty orators who have vast audiences hanging on their every word. Make your concept as potent as it can possibly be.

Your concept should be Intelligent and Emotive

Your idea should appeal to your audience's hearts and minds. So be wary of concepts that are cold, intellectual exercises lacking in emotion (although sometimes the emotion in a concept can be the warm feeling people get from understanding your intelligent reference).

Consider your concept from both angles, and if it's lacking a rational or emotional aspect, work out how to add whichever is lacking to the idea. Just consider what missing *thought* or *feeling* you could bring in to your concept.

Essential example

The Nike 'Write the future' TV ad had a simple, original concept with an idea that appealed to people's minds: inventively imagining what famous footballers' future lives might be like after doing well at the World Cup. But the concept was also rich with emotion, with moving music, slo-mo action and cheering crowds that stirred the blood and made you want to be part of it (by buying Nike, obviously).

So that's SOPHIE. Remember her three couplets and you'll know what to look for in your concepts. Now, how to come up with them. Enter stage right: BOB.

BOB

'Oh, for a muse of fire that would ascend the brightest heaven of invention!' begins Shakespeare's *Henry V*.

Sometimes, after a long day and with a deadline pressing, a muse of any kind can be hard to come by.

That's why over the years I've gathered, developed and post-rationalised lots of ways of developing concepts. Together, they make up BOB – the Book Of Brainwaves. Using it helps prevent you from getting into a rut of coming up with the same sort of ideas over and over again.

I'm going to briefly cover ten of them here – and of course you won't get a chance to explore all ten for every communication. But it is worth looking at two or three each time (and varying at least one of them each time) to come up with a potent idea.

Then, to select the best concept, evaluate them with SOPHIE. And also consider which best suit the medium you're writing in and the audience you're writing to.

The Book of Brainwaves

1. Dramatise
2. Factualise
3. Reframe
4. Genre
5. Challenge convention
6. Topical
7. Analogy
8. Perspective shift
9. Become another
10. Anecdote

Dramatise

A simple one to start. Just exaggerate and dramatise the proposition until it becomes creatively compelling.

Since your proposition is usually around the key benefit, message or solution to a problem, dramatising that benefit will nearly always give you a strong concept for your writing. So this approach is always top of my list.

Essential example

An idea we came up with for Husqvarna, who make those sit-on lawnmowers for rich people with big gardens. One of the key benefits was that large lawns could be mowed quickly and easily. To dramatise that to an extreme: what if it made mowing your lawn so quick, it was as if your lawn felt tiny to mow? So we designed a one-piece mailing stuck to an A5-sized piece of astro turf, with the line 'Makes mowing so easy, it's as if your lawn's this big.'

Factualise

Find a fact, figure or stat and lead with that. A surprising fact can be very effective, because it can be creatively engaging and utterly truthful. The stat may be directly connected with your subject, or related to it more indirectly, if it helps illustrate your message.

Essential example

The poster for Cancer Research UK that shows three little girls sat with their backs to us, looking out over some fields. There's one word above each of them: 'Lawyer, Teacher, Cancer', highlighting the fact that one in three people will be affected by cancer.

Or the WWF mailing – the letter headline says 'This letter contains 300 words – one for every snow leopard left in Nepal.'

The other way you can use this technique is to discover a fact (rather than a stat) about your subject, and then lead with that.

Reframe

Just as you might rewrite a headline a dozen different ways, try doing the same with your proposition.

Change the words, write it so it says basically the same thing but in another way, imagine how different people might express the same thought ... just find as many ways as you can to rephrase and reframe the proposition. It'll often give you a new way to look at the core message that could form the basis of a new concept.

Essential example

The original proposition for British Airways was 'The world's biggest airline'. But it's clear how much more interesting they could be when you simply rephrase that proposition as 'The world's *favourite* airline'.

Genre

Choose a storytelling genre and see how you could bring your subject to life with that theme.

Most people grow up with stories from a very early age. And people continue to be drawn to stories – novels and films, for instance. So presenting the idea as some kind of story can be very engaging.

Genres you might consider include:

- horror
- thriller
- adventure
- comedy
- action
- fairytale/fable
- love story
- period drama
- soap opera
- documentary
- news
- chat show
- reality show.

Alternatively, think about how you might tell a story differently according to whether it was the story in a film, play, musical, graphic novel or book.

Challenge convention

Ask yourself, 'What if we *couldn't* do it that way? How else could it be done?'

Challenging the convention is good for coming up with bold ideas

– and the bigger the convention you challenge, the bigger the idea you'll get.

It's useful because we make a lot of assumptions in almost everything we do. Usually those assumptions are handy shortcuts for getting things done, but they can mean we end up doing the same things in the same way, again and again – which isn't very engaging.

Put the conventions/assumptions you've been working with under the microscope and ask why – *why* does it have to be done that way?

Essential example

If you're writing a letter . . . does it have to read top to bottom? Does it have to be about the subject? Does it have to be signed by you, or could someone completely different sign it? Elvis, perhaps, or you ten years from now, reflecting on the company's success over the last decade since they took the action you proposed? Does the letter have to be in English? What if you got the first half translated into Latin?

An effective alternative to asking *why* is to create an artificial problem to get around.

So you might say, 'Headlines must only be one word long' or 'The presentation has to look like a six-year-old did it' or 'I'm not allowed to mention the product name'. Necessity is the mother of invention – so create a false necessity to come up with an inventive new approach.

Essential example

The iPhone is full of ideas which challenged conventions of the time. As a starter, the keyboard. Posing it as a problem, they could have said, 'How do we make a phone . . . when we're *not allowed* buttons?'

This could have led to the touch screen – one of the benefits being that when the keys aren't being used there's more area of the phone's 'real estate' given over to the screen. Another being that since the buttons are virtual, it's easy to customise and rearrange them.

Topical

Christmas, New Year, Valentine's Day, Spring, the anniversary of the subject, on this day in history – all ways of being topical that can drive a concept. But also look at what's in the news and what the current Zeitgeist is. What's likely to be going on in the lives of our audience? What stories are grabbing their attention?

Essential example

On the day George 'Dubya' Bush left the White House, under an article about his departure, Veet (hair removal cream for women) ran a small press ad: 'Goodbye Bush'.

Analogy

Metaphors, similes and analogies – just look for something that is, in some way, like your subject (like the proposition of your subject).

Essential example

Volvo want to get across that their cars are safe, so they show a safety pin bent into a car shape.

Land Rover want to convey that their cars are tough and well-suited to the outdoors. So they show a line of Masai tribesmen of different heights to match the car shape, holding shields to represent the wheels.

Do sense-check them with other people though – it's easy to come up with an analogy that you think is great, but which other people just don't get. Analogies are a form of 'borrowed interest'. In the Volvo example above, for instance, it's borrowing the reputation a safety pin has for being safe. Which is fine; just be sure that what you're borrowing is obvious to your audience.

And generally it's better to find something in your subject that's interesting, rather than having to borrow the interest from elsewhere.

Perspective shift

This simply means taking a different point of view. You could look at the subject through the eyes of someone unexpected, or change the time, to look at it from the past or the future.

Essential example

The Gü chocolate brand. Do you know this story? The version I've heard is that there was a businessman who loved chocolate and decided to set up a chocolate dessert business. He briefed an agency to work on the brand . . . and after a few weeks, hadn't heard a dicky bird.

So he called them up. 'Oh', they said, 'that. Yes, well . . . you'd better come in.' So he did. And they told him, rather sadly, that they were too late. When they were doing their research, they'd discovered that another chocolate dessert brand had just launched, and it was called Gü. They showed him the luxurious black and brown packaging. The great language. Even the name 'Gü', how it sounded, how the umlaut over the 'u' made it look like a smiley.

The ad agency wasn't sure he'd want to compete against this new entrant, because the Gü stuff looked pretty good.

'Pretty good?!?' said the businessman. 'This stuff is brilliant! What a disaster – if I'd had this brand I could have been incredibly successful!'

At that point the team at the ad agency paused. And smiled. 'Well', they said, 'if you really feel like that, then we've got some good news. Because Gü doesn't really exist yet. It's the idea we've created for you.'

What a brilliant concept: to shift the perspective and imagine if they weren't showing their client work for them, but work that had already been bought by a competitor. It would make their client want the work all the more – and be delighted when it turned out they could have it.

And of course, Gü became a very successful brand. And I believe the businessman who had started it, for £100,000, went on to sell it for £35 million.

Become another

Imagine how a completely different brand would tackle your subject. Or a famous person. Become them. Think like them. Do an idea like them.

Essential example

How would Honda write your communication? How would DFS do it? Or Marks and Spencer? Or maybe, what if it had been written by Stephen Fry? Or Sly Stallone? Or Kate Moss? Or if it's a poster, imagine tackling it as a website brief. Or as a packaging brief.

Anecdote

Loved by stand-up comedians and after-dinner speakers, anecdotes can also work well as the concept behind your business writing. Back in *Perspective shift*, for example, I told you the story of Gü. It's not my anecdote, it doesn't have to be. It's just a 'true story' that illustrates the point.

If you have an anecdote that's relevant to you, use it. But if it's just a story you've heard elsewhere, and it's interesting and has a natural fit with your message, then that's fine too.

Essential example

In a blog post I wrote, I was talking about the march of technology and how lots of young people don't wear watches any more, because the time's on their phone. I began with an anecdote of my mum's: of how when she was a girl, young people didn't wear watches either – not because they didn't want them, but because they couldn't afford them.

In fact, in the summer, they'd cut pieces of paper in the shape of a watch, and put it on their wrist when they were out in the sun. That would give them a 'watch tan-line', so the rest of the time it would look like they did own a watch . . . they just didn't happen to be wearing it at that moment.

And there you are – ten concept-generation ideas. As you've probably realised, they can be useful for more than just concepts for business writing. They can be invaluable when brainstorming anything, from product development to process innovation.

6

Essential Plan checklist

Here we are then. At the end of our planning and ready for action. Ready to create powerful business writing that tells a simple, engaging, persuasive story. But first, let's review the five planning steps you should take before beginning any significant business writing task:

1. Remember that you'll be aiming to write in a way that:
   - feels easy, simple and natural for the audience to read, understand and agree with;
   - tells a story which keeps the audience utterly engaged throughout;
   - makes them feel glad they read it;
   - persuades them to think/feel/do whatever it is you want from them.
2. Formalise a proposition; a compelling, single-minded idea that your whole communication will be centred around.
3. Gather the best content to build the case for that proposition, through interrogation and insight.
4. Consider the context of your communication – the audience you're writing for and the medium you're writing in.
5. Develop a concept to bring your story to life, the way an ad agency would if you'd briefed the job to them.

And let's recap on ten of the most common mishaps that can mangle your business writing:

1. Thinking a complex product/service/idea needs a complex explanation.
2. Writing about too many things at once.
3. Starting slowly and continuing vaguely.
4. Writing to impress, rather than to communicate.
5. Writing for yourself, not the audience.
6. Ignoring the advantages and limitations of your medium.
7. Forgetting to create communications which are as appealing to look at as they are to read.
8. Writing copy with too much tone/too little tone.
9. Being only rational and forgetting the emotion.
10. Thinking you should simply inform or entertain, rather than persuade.

I would recommend that, at this point, you create an 'executive summary' page for yourself: the distillation of your planning that's going to be an *aide-mémoire* for when you write your first draft.

And on that summary you'll write:

- Proposition
- Insight
- Nuggets
- Audience
- Concept.

Under each, write a summary of what you've gleaned from your planning. Under 'nuggets', for instance, you'll write three, four or five interesting facts/findings/statistics/case studies that have come out of your interrogation.

The planning process is an important one. Every painter and decorator knows that time spent preparing the surfaces properly will save them time later and – more importantly – will give a better result.

It's exactly the same with your business writing – and just like a painter and decorator (if you can live with the analogy) you'll get quicker with practice. The planning stage will gradually evolve from a laborious, time-consuming, back-and-forth series of steps into a seamless part of your writing and a smooth sense-check of your approach.

So. You've a pile of notes. You've a hatful of ideas and a fistful of facts. You've given yourself the 7.62mm full-metal-jacket, armour-piercing ammunition to get your point across … and you know who you're aiming at.

Now it's time to pull the trigger.

part

Doing

Here's a happy coincidence: the proportions of this book roughly match the proportion of time you should consider spending on each when developing your work. In other words, a third of your time on the *planning*, a bit more than half on the *doing* and the remaining … sixth … on *reviewing*.

So this is the major part – the *doing*. I've split it into two main acts: the first three chapters cover the *doing* fundamentals, while the next three – winning hearts and minds, being fascinating and being persuasive – cover more advanced *doing* concepts.

Then there's a chapter of 25 'quick wins': simple tricks you can use at a moment's notice to turn up the volume on your communication's potency.

So, pick up the Mont Blanc you got for your birthday or the free ballpoint you took from that convention. Or, more likely nowadays, select *File, New* on your screen and we'll begin with the *doing* basics.

7

Punctuation, grammar & usage

I apologise: looking at punctuation is of course extremely basic.

However, most of us haven't had an English lesson in 10, 20 or more years, so a quick refresher of the most commonly made mistakes in business writing may be useful.

You may find some stuff you didn't know. Or things you once knew, but have perhaps forgotten. Or have just never been 100 per cent certain about.

Either way, it can be embarrassing to get the basics wrong. It strips the professional veneer from your writing and might make your reader think, 'Do they really know what they're talking about? I mean, they don't even know how to use an apostrophe!'.

So, if words are the bricks you're using to build your case, you should at least be confident that you know how to lay them properly.

Punctuation & grammar

English is an evolving language.

We might not always like the way it evolves (I hate the way people say 'real' instead of 'really', for example).

But the alternative is to not evolve. Like French. French is carefully regulated (by things like the Toubon Law and organisations such as

L'Académie Française) and has a famously small vocabulary (excluding scientific and technical words, perhaps just 50,000 to our 250,000).

The trouble is, if you don't evolve, you get left behind and end up with *nil points*. So we have to get used to the constant mutation of the English language; new words like *screenager* (a teenager who lives online) appear, old ones like *egrote* (to feign an illness) die.

Punctuation and grammar change too, making it difficult to be definitive about what's right and wrong. But what we can agree on is the *purpose* of these things.

And it's still to communicate. As effortlessly as possible. Which means three things.

One, we should know the current, official guidelines for punctuation, grammar and spelling.

Secondly, we should bear in mind our audience.

If your audience is older and educated, they may not like split infinitives or sentences that begin with a proposition. In which case, don't do it – because powerful business writing should get your audience on your side, not alienate them. However, most audiences won't mind (or even know about) such things.

Thirdly, good writing is often a little conversational in tone. And people don't speak in perfectly arranged sentences – they play fast and loose with the rules of usage, frequently using sentences that don't contain a verb, for example.

Fine. Your writing can too (like the preceding one-word, verb-free sentence, for example). It makes your writing more approachable, more engaging and easier to absorb.

In other words, spelling and punctuation you should *always* do correctly. Grammar you should *usually* do correctly. Usage ... well, that's a more moveable feast.

These are some of the common mistakes in spelling, punctuation and grammar – including some that your spell-checker won't pick up:

- *'It's' and 'its'*: I'm starting with this one for two reasons. First, because the correct usage is really easy to explain and remember. And secondly, because it gets used incorrectly *everywhere*. I've seen it wrong on BBC television news captions, for instance. On film subtitles. In magazine articles. And on scores of business documents.

  Here's the answer: 'It's' is only ever used when it's short for 'it is' or 'it has'. Otherwise you use 'its'. No need to think about possession or anything complicated like that. Just read the sentence, substituting the occurrence of 'its' with 'it is' (or 'it has'). If it makes sense with 'it is' (or 'it has'), then you use 'it's'. If it doesn't make sense, use 'its'. And that's all there is to it.

- *'Your' and 'you're'*: again, very simple. 'You're' is short for 'you are', so just see if the sentence makes sense with 'you are', in which case use 'you're'. If not, use 'your'.
- *Could of/would of/should of*: this is as a result of people writing what they hear. People say 'could have' in a way that sounds like 'could of'. So you see 'could of/would of/should of', when the correct phrase is 'could have/would have/should have'.
- *'Literal' for 'metaphorical'*: I've seen documents with phrases like 'It literally blew my mind', when of course they mean the exact opposite – it *metaphorically* blew their mind. If it had literally blown their minds, they wouldn't be capable of writing about it.
- *'Their' for 'there' or 'they're'*: 'their' means something that belongs to someone (like 'their socks'), 'there' is a place (like 'you left your socks over there') and 'they're' is just short for 'they are' (like 'those socks of yours, they're disgusting').
- *'Maybe' and 'everyday'*: 'maybe' and 'may be' are not interchangeable. Substitute 'maybe' for 'perhaps' – if it makes sense, then OK. If not, then you mean 'may be'. For instance, 'It may be right to change this, what do you think?'. 'Maybe.'

  Similarly, 'everyday' means something is ordinary and commonplace. So if you take a vitamin tablet each morning, you're taking one 'every day', not 'everyday'.
- *DVD's:* when you're talking about your company's new range of DVDs, you don't need the apostrophe. Or after TVs or 1000s or the 1990s, any time you're just talking about the plural of something that has capital letters or numbers. However, using an apostrophe here isn't the end of the world, and it's often done because sometimes it can make things clearer – for example, when the last letter is a vowel.
- *'Desert' for 'dessert'*: I saw this on a menu in a Lake District restaurant recently. I wanted to order one, but I was worried it might be a bit dry.
- *'Loose' for 'lose'*: You don't 'loose' a game of golf, you 'lose' it. Unless you win, of course.
- *'Lead' for 'led'*: in the present tense, you 'lead' someone. In the past tense, you 'led' them. Yet 'lead' is often mistakenly used

as the past tense form – perhaps because the metal 'lead' is pronounced 'led'. But when you're talking about leadership, you say, 'When we went to Machu Picchu last year, I led the expedition.'

- *Parentheses*: or brackets, if you want to be common. If the open bracket starts during the sentence, then the full stop will finish outside the close bracket (like this). (However, if the open bracket starts before the first word, the full stop is inside the closing bracket, like this.)
- *Less or fewer*: less means 'not as much'. Fewer means 'not as many'. So because we're talking about 'not as many people' rather than 'not as much people', the correct usage is 'fewer people', not 'less people'. We say 'less than two weeks' because 'two weeks' is how much time we're talking about, not 'how many time'.
- *Use of 'a' or 'an' before 'h' words*: some would-be sticklers insist that words beginning with an 'h' should be preceded by 'an' not 'a', as in 'an hotel' or 'an historic occasion'. But actually, if the 'h' is pronounced, use 'a'. Use 'an' where the 'h' isn't pronounced. So it's 'an hour', but 'a hotel'.
- *Hopefully*: like most people nowadays, I use hopefully 'incorrectly'. I do know the correct usage though: to do something hopefully means to do it positively. Hopefully does not mean 'fingers crossed'.

  However, this is where grammar becomes rather muddy: hopefully has been used in a 'Hopefully we'll get to the station in time' kind of way for a hundred years, but some people still object to it. So, if you're cultivating a pedantic personality, only use hopefully in a 'We will pitch for the business hopefully and energetically' kind of way.
- *Momentarily*: the UK usage means *for* a moment, as in 'We all paused momentarily before agreeing'. The US usage means *in* a moment, as in 'I'll be there momentarily' – I'll be there soon, not I'll be there for a very short period of time.
- *Proper nouns*: proper nouns are things that there are only one of. The London Stock Exchange, for example, is a proper noun, so has capital letters. Blackpool Tower, the same. There are also intangible proper nouns, such as the Pareto Principle, for instance.

  The clue is the use of 'the' at the start, the definite article letting us know that the noun we're referring to is unique.

But in Business Speak Everything is apparently So Important that People start giving Capital Letters to almost every Noun they Write.

So a bank refers to their Bank Accounts. A fast-food chain talks about its Customers. And a technology company refers to its new Server Systems. It makes your writing look German (where every noun has a capital letter) and it feels pretentiously grandiose – as if your bank account is something of unique importance in everyone's lives. Don't Do It.

- *Single entities*: your company is a single entity. Like Shell, for instance. So, Shell should say 'Shell *is* proud to announce ... ' not 'Shell *are* proud to announce ... '. Or 'In the last 12 months, B&Q *has* opened 12 new stores in the UK', not 'In the last 12 months B&Q *have* opened 12 new stores in the UK'. So many business communications get this wrong; yours doesn't have to be one of them.

OK, that's enough on the basics of punctuation and grammar. The point with this stuff is to know what's important and what's not, how to avoid silly errors and how to be mindful of what will rub your audience up the wrong way.

There are infinite resources on the Web; it never takes more than a few moments to find the answer to any spelling, punctuation or grammar query you have. Get a good grip of things like apostrophes (such as using with plurals, like 'two weeks' notice' and you'll be fine.

For books on punctuation and grammar, some that I've read and found useful are John Humphrys's *Lost For Words*, R.L. Trask's *The Penguin Guide to Punctuation* and Lynne Truss's *Eats, Shoots and Leaves*.

Essential example

If you google (another word that's only joined the dictionary in recent years) Lynne Truss, you'll find a number of people who claim that there are a fair few punctuation and grammar mistakes in her book.

For instance, the subhead to the book is 'The Zero Tolerance Approach to Punctuation'.

Even I know that neither 'zero' nor 'tolerance' can work independently as adverbs to 'approach', so they should be hyphenated: zero-tolerance. Truss, a self-confessed super-stickler, gets her punctuation wrong ... on the front cover of her book about punctuation. Which just goes to show what a minefield it can all be.

Usage

Have you heard the (possibly apocryphal) legend of the NASA Space Pen? NASA needed its astronauts to be able to take notes and make reports while up in space. And Bic Biros don't work in zero gravity.

So NASA got to work. Some of the biggest brains on the planet spent 10 years and 10 million bucks solving the problem. And they came up with the 'Space Pen', which does indeed work in zero gravity, as well as underwater and at any angle.

Russian cosmonauts, of course, had the same problem. But Russia came up with a somewhat simpler solution. They gave them a pencil.

Once again, I'm referring to the benefits of simplicity. In *Planning*, we looked at the importance of expressing your ideas simply (but not dully). That extends right to the words you choose and the sentences you construct.

When it comes to usage, don't use more words than you need to, or more complicated phrases than you need to. As Mark Twain said, 'Eschew surplusage'.

Essential example

'Pre-warn'. It's become a commonplace phrase in business, but how is it different to 'warn'? If I pre-warn you about an emergency meeting, does that mean that later I'm going to 'warn' you, about the same emergency meeting?

Another one is 'price point'. How is a product's price point different from its price? I've never found anyone who can answer that satisfactorily, yet every FMCG (fast-moving consumer goods) business refers to its products' 'price points'. Or 'media forms'. Media is the plural of medium, so of course it's more than one 'form'. You can just say 'media'.

Double negatives or indeed any needlessly negative constructions are another easy complication to remove.

Essential example

Here's an email I received from my local Apple reseller: 'As your local experts in all things Apple, we have not only the full range of Apple products and accessories, but also the expertise to help you . . .'.

The 'not' isn't supposed to be a negative – they're 'not' talking about something they lack. So instead of saying 'not only/but also', it would be a simpler construction (and quicker to the benefit) to say 'As your local experts in all things Apple, we have the full range of Apple products and accessories and also the expertise to help you . . .'.

Just a small change, but simpler and stronger. Although there's more wrong with the paragraph than just the negative construction – for instance, they're saying that because they're 'experts', they have 'expertise'. Well, duh.

Look for simplicity in every aspect of your writing – from simple things like keeping the vast majority of your sentences under 30 words long – to the way you phrase your story.

Write in a friendly, easy style and use simple, short words wherever possible. Always look for the shorter, simpler word. Look for words with fewer syllables. Look for words which are more 'concrete' (rather than abstract) and more common (more Anglo-Saxon than Latin). The occasional unusual word can really work well, but that's the *occasional* word. The other 99 per cent should be the easiest to read you can find.

The best way to explain what I mean is with another example. Here's the opening to a business email I received the other day (I've hidden the company name):

> *I'm writing to let you know about an exciting new Group programme that is underway – the [company name] flagship social programme. See attached for a little more information about and examples of other successful flagship social programmes. This will be a bespoke programme that gets our people involved in creating social value in a way that makes strategic sense for us as a business and utilises our skills, strengths and resources. It is intended to complement rather than replace any corporate responsibility or charitable initiatives currently in place.*

Does that sound like simple usage to you? It's exactly the sort of writing-without-communicating that we want to avoid.

It uses complex constructions ('information about and examples of'). It mixes 'normal usage' English with email shorthand English ('see attached' instead of 'see the attached file'). It repeats the word 'programme' four times in three consecutive sentences. It starts with 'I'm writing', a terrible opening (I know you're writing, just as I know I'm reading).

And worst of all, underneath all this needless complexity ... it doesn't really tell you much at all. I have little idea of what it's really about or why I should care. It could have been captured more simply – and more elegantly, powerfully and emotively, like this:

> *We've got some great people at [company name], with highly prized skills. Skills we'd like to harness in a new, flagship social programme.*
>
> *Social programmes are not commercial ventures. Instead, they're projects which use our people's experience and expertise to make a positive difference to society.*
>
> *We've developed a new social programme to give more people at [company name] the chance to get involved and make a contribution. You can read a little more about what this means in the attached proposal – which includes an example of 'Positive Parenting' – one of our first projects in action.*

It's actually a little longer. But the average sentence length has dropped from 22 words to less than 16, so it feels easier to read. We've ditched words like 'utilises', 'initiatives' and 'strategic sense'. And we've broken up the paragraphs, so again it *feels* simpler.

And, even though it's only eight words longer, it has more emotion and energy to it. The voice is more 'active'. It explains what a 'social programme' is more clearly and it names a specific example of one for you to look at in the attached document.

In other words, writing more simply enabled me to communicate more effectively.

Essential tip

When you've written your first draft, always read it out loud. And really listen to what you're saying. Does it sound natural, human, conversational? If not, make it so. Better yet, get someone else to read it out loud to you – because then you'll see if they read it with the rhythm, inflection and meaning you intended.

Typography

We covered the basics of typographic legibility in *Planning*. Here are a few more useful pointers.

In print – especially long writing – serif typefaces are more easily read than sans-serif ones (which is why magazines, newspapers and books like this one use them). The reason is, our eyes don't take in individual letters, they recognise word shapes. Serifs (by which I mean the little sticks on the ends of the letters, such as those on these words) make those shapes quicker to recognise.

Online, where writing tends to be shorter anyway, sans-serif fonts are often more legible (because screens are made up of pixels that sometimes don't cope with the detail of serifs very well, creating unclear shapes).

Having said that, sans-serif fonts seem to have taken over *everywhere* in the last decade – perhaps because Arial was the default font on so much software – so you'll see it in lots of printed long business writing.

Apart from serif or sans-serif, the other two key considerations for legibility are colour and size. For colour, the starting point is: do you have black text on a white background or white text reversed out of a dark-coloured background?

Black text on white is much more legible in print than reversed-out text. Reversed out is fine for headlines and short writing, but harder to read for long text. And be very wary of putting your words over a picture – it will be very hard to read, so most people won't bother.

A current trend is light-grey text on a white background (especially online) – again, avoid as it's hard to read. Unless what you've written isn't worth reading. Online, reversed-out text – in a large enough sans-serif font – can actually work OK, as a dark background means there's less backlighting, so less eye strain.

Essential tip

As well as the size of your font, look at the leading (pronounced 'ledding'). That's the size of the font that's used to determine the distance between each line of copy. For instance, '11 on 10' means the font is 11 point, but set in a line grid of 10 point, so it'll be very close together; '11 on 12' means it's still an 11-point font, but set to a 12-point grid so there'll be a wider gap between each line and it'll be easier to read.

And remember: bigger is not always better. Too big a font or too much leading can make your text look clunky and disjointed and actually harder to take in – like sitting in the front row at the cinema. Making a 12-word heading so big it goes onto five lines is going to be more difficult to read than having it smaller over three lines.

Look for a size that is both pleasing to the eye and easy on the eye. For A4 documents, 11 point is usually a good size for body text.

Font choice

Fonts convey feeling and meaning. So the font you use can have an impact on your communication – choose a font that suits the tone, and the feelings will be amplified.

The opposite is also true. For instance, I've worked for a number of charity clients that need emotive, moving writing to persuade the audience to make a donation.

Yet their 'house style' insists on using Helvetica for body copy. Helvetica is a cool, calm font, devoid of tension or emotion. A more dramatic-looking serif font would be much better to convey the story (and much more suitable for an older, charitable audience who are more accustomed to serif fonts anyway).

Fonts are the clothes your words wear. So get to know them, see how they look at different sizes and styles, and don't be afraid to mix a serif and sans-serif font in the same piece, provided they sit together well.

Line breaks

I think it's useful to consider the proportions of important lines of text, such as headlines. Namely, where you 'break' a line, if you need to. For instance:

Is this headline OK?

only really works if you can get it all on one line. There's nowhere neat to break it – you'd have to have:

Is this

headline OK?

or

Is this headline

OK?

which is even worse. So if 20 characters in the headline still means the font is big enough, then:

Is this headline OK?

is fine. But if you decide that 12 characters across a headline is the most you'd really want then you may want to re-write it to:

How's this

headline?

perhaps. Which balances much better – the first line is 10 characters long, the second 9.

I try and get my headlines to break at a point that not only balances them well physically, but which doesn't break the line in the middle of a phrase – for instance:

Get a free

gift today

balances very well in terms of length, but it splits the phrase 'free gift', which makes it more awkward to read. Instead:

Claim your

free gift

breaks more logically.

Spending time considering these things will make it more likely that your words get read. And since you spent all that time choosing them, isn't that a good thing?

Dashes and hyphens

A tiny typographic point (or, rather, line) to finish on.

Dashes between phrases should be 'long dashes', known as 'em-space' dashes (because they are the length of the letter 'm' in the same font and size).

Type a dash on your keyboard and it first appears as an 'en-space' dash (ie, the length of an 'n'). It may auto-correct to an em-length dash when you then press the space-bar. See how the phrases 'auto-correct' and 'space-bar' are linked by hyphens (which are en-space in length), whereas phrases – like this one – are separated by dashes (which are em-space in length).

OK, that's the basics of punctuation, grammar and usage covered. Onwards and upwards.

How to harness style & structure

Style

I should begin by pointing out that your writing already has a style, whether intentionally or not, just as you have an idiolect (your individual speech habits). However, the way people write is rarely the same voice they speak with. In fact, it's often a voice *no-one* would speak with.

Perhaps you know someone with what used to be called a 'telephone manner'. They're talking to you perfectly normally, then they're introduced to someone they've never met before. And suddenly they're all, 'Oh, hel-low, how orfully dee-lightful to meeeet you, reeeelly sow ek-citing'. They try to sound all educated, like what Mrs Malaprop did.

People often adopt their own special tone of voice in business writing too. Instead of trying to write the way they'd speak, they write in an affected, stilted and pseudo-clever way.

Instead, aim for a style that is natural and conversational in tone – try and make it sound *more* like your speaking voice, not less. It sounds more personable and avoids creating distance between the writer and reader, which formal language does.

With that in mind, I would say that the discernible style of any business writing is a heady mix of the subject, the concept, the tone of voice you're aiming for … and some indefinable bit of inspiration that hits you while you're writing it.

That can be one of the pleasures of powerful business writing. For all your *planning*, for all that your experience and expertise guide you to what's 'right', every communication still grows in its own way. You're the gardener, tending and encouraging it, but you still can't predict exactly how it will turn out. To help cultivate the style you're aiming for, here are two 'personality penmanship' techniques you might find useful.

1 Who am I?

One way to write with a particular style is to think of someone you know, or a famous person whose voice you can call to mind (or watch clips of on YouTube) to imitate. It just gives you a slightly different perspective from writing as yourself.

Professional ad writers do it all the time of course, since they write on behalf of brands. So they can only use the tone of voice and personality that suits that brand.

Essential example

Once I had to write a letter that was being signed by Lady Thatcher. So not only was I writing about her life and experiences, I was writing it in her voice, too. One sentence was: 'And on the 4th May, 1979, I became your Prime Minister'.

If I'd been writing it in my tone of voice, I'd have said 'And in 1979 I became Prime Minister'. I wouldn't have put the specific day, but I thought she would be very particular about it.

I wouldn't have said 'your Prime Minister' either. I'd have been squeamish about saying 'your', worrying it sounded rather self-important and assumptive (especially as not everyone voted Conservative). But again, I thought it was the kind of language she would use.

On reflection, maybe I should have said 'And on the 4th May, 1979, *We* became your Prime Minister'. The Iron Lady was fond of using 'the royal we'.

2 Personable adjective

So you can cultivate a style by imagining a person. Alternatively, you could keep in mind an adjective which best summarises the tone you believe would be most effective. Here are some examples:

- Authoritative
- Charming
- Witty
- Hushed
- Conspiratorial
- Breathless
- Enthusiastic
- Angry
- Seductive
- Worried
- Reassuring
- Mysterious
- Calm
- Soothing
- Friendly.

Essential tip

There are many ways to give your writing a little more style, but here's something that doesn't: an avalanche of adjectives and adverbs.

Sometimes people think that to make their writing more interesting, they should smother every sentence with descriptive adjectives and adverbs. Actually, it makes your writing sound like bad romantic fiction. Ruthlessly edit out these modifiers and you'll often end up with a much stronger, punchier communication.

Do be wary of trying to imbue your writing with too much style or personality. First, it can be hard to do well; secondly, it can overwhelm the content; and thirdly, it can be misinterpreted. That's because a) you're not able to judge the audience's reaction to see if they like your tone, and b) the written word can make personality more ambiguous than the spoken word.

Essential example

'And of course, who wouldn't want to spend the day exploring the nearby salt museum?' If that was spoken, you'd be able to tell that the speaker was being sarcastic. Written, who knows? Maybe they find salt fascinating.

Ultimately, you're looking for a style that you believe will create the strongest connection between the two of you.

Michelangelo put it well. He said that he didn't create statues like *David*. They were already there, in the stone. He simply chipped away the bits that weren't needed, to set them free.

Structure

There's a Morecambe and Wise sketch where conductor André Previn says despairingly to Eric Morecambe (playing the piano in Previn's orchestra), 'You're playing *all* the wrong notes'.

Eric peers at him (irritably) through those thick-framed NHS glasses. There's a pause. 'I'm playing all the *right* notes', he snarls. Another pause. 'Just not *necessarily* in the right order.'

Structure's like that. It's there to make sure that what you write is in the right order for maximum impact and effectiveness. Now, as I'll explain in a moment, there is no universal 'perfect structure' for powerful business writing. Every piece is unique, and there are dozens of perfectly good structural solutions for each piece.

That said, there are some basics that are worth observing. Including some wisdom straight from Aristotle. Back in … olden times, Aristotle came up with a four-point structure for rhetoric. It goes:

1. *Exordium* – Make a shocking statement.
2. *Narratio* – State the problem.
3. *Confirmatio* – Offer a solution.
4. *Peroratio* – Explain the solution's benefits.

The more modern iteration which you'll almost certainly know is AIDA. Attention, Interest, Desire, Action, known as the 'purchase funnel'.

It's the idea that you should first get your audience's attention, then tell them something about your product/service that will interest them, create a desire for what you have to offer, then get them to take action. And it's not a bad starting point for structure. Even though you might not have separate, distinct sections of writing that are *just* about Attention or Interest or Desire.

So, will your copy grab attention, right from the get-go? Will it be interesting in both style and substance? Will it create genuine desire for the product/service/proposal/recommendation/investment you're trying to sell them? And, if there is an action, are you making that action as clear and definitive and important as possible?

Attention, for example. You really need impact to get an audience's attention. Always remember just how busy your reader is. How many demands are being made on their time. How many letters, emails, presentations, memos, reports, proposals and so on they get bombarded with.

And, in fact, you're competing with *everything* that's trying to get your audience's attention. The dog being sick. That red reminder from the phone company. A new drama on TV. A mild hangover. A major love interest.

Against that onslaught of demands, what chance have your carefully crafted words got? Not much, unless you can grab your audience's attention in the hint of a shaving of an iota of a scintilla of a soupçon of the nanosecond that they glance in the direction of your communication.

However, a word of caution. Nowadays, people talk about 'relevant abruption' or variations thereof. It's a qualified, more sophisticated version of the 'Attention' part of AIDA. It means, get your audience's attention *in a way that's relevant* to them and your subject.

After all, you could just have the headline FREE MONEY. And that would probably get attention. But if the communication isn't really about free money, then you're not getting their attention in a relevant way. They'll read on long enough to realise you've misled them.

At that point they won't carry on, despite your best hopes. They'll turn away, annoyed and, like the boy who cried wolf, you'll have tricked them to your own disadvantage: they'll be less likely to pay attention to you next time. Even if you then have something that would genuinely interest them. So get your audience's attention with 'relevant abruption' before drawing them in further … and arousing their desire.

AIDA is only one way of creating a basic structure. Another is to use a headings template. For instance: *executive summary, introduction, background, proposal, evidence, swot (strengths, weaknesses, opportunities, threats), conclusion, next steps.*

Personally, I don't put much stock by that kind of approach – and it's often a surefire way to stifle your writing. It might seem like a logical method for organising your thoughts, but it's also a very staid, one-size-fits-all approach.

Here's an alternative, more organic way to develop a structure; I (rather grandly) call it the *Song Method.*

Essential tip

Think of your communication as a series of segments. It's those segments you want to move around in order to create the strongest structure.

Begin by writing your first draft, quickly and roughly, in whatever way comes naturally. Then go back and identify the different segments of your writing.

If you're writing something very short you might not have many segments – in which case structure is probably irrelevant anyway. It doesn't play a big part in billboards, for instance. But aside from that, you should be able to segment your communication quite easily. The 'opening that builds on the headline' segment. The 'metaphor' segment. The 'main benefit brought to life' segment. The 'call to action' segment. The 'reprise of the opening' segment, the 'first example' and so on.

Turn this into a list. A list of segments, each of which might vary wildly in length, which is absolutely fine.

Think of these segments as being like the parts of a song. A song, for instance, might have an intro, three verses, a bridge, a chorus, a guitar solo, a version of the chorus that goes up a key, a 'middle eight', an outro. And you could arrange those parts in many different ways.

'Intro, verse, bridge, chorus, verse, bridge, chorus, guitar solo, verse, bridge, enhanced chorus, outro.' That might be a typical kind of song structure. But it doesn't have to be like that. Plenty of good songs start with the chorus. Or have three verses before the first chorus. Many don't have a guitar solo, of course.

Now that you have your own segments, try moving them around to see which order makes the most powerful 'song'.

Obviously you're bearing in mind all of your *planning* – so what you know about the audience, the medium, what your proposition is and so on. And you'll be using everything we've yet to cover – how to be fascinating, how to win hearts and minds, psychological persuaders – to guide you. But even so, you'll find you can move things around to create different but equally effective structures, according to the feel you're going for (to stick with the song analogy, whether you feel your piece should be a ballad, an opera or a rock tune).

Essential tip

I often have two or three documents open at once, all with the same piece. But in one I'll have the words in the original structure, in the others I'll try a couple of variations with the structure, moving the segments around in different ways. Most times, it's one of these I'll end up using, rather than the more 'traditional' structure of the original.

While you're experimenting with structure, here are eight useful pointers:

1 Don't have an 'introductory segment'. You'll have a first segment of course, but don't think of it like an introduction or you'll write something too soft and slow-paced.

2 Let the segments build on each other. A segment should move on from the previous segment, not a) jump around in time or b) seem completely unrelated. These are not a series of distinct subjects after all. This is one story divided into chapters.

3 Don't retrace your steps. (Apart from when you're deliberately making a reference from one segment to another.) Otherwise, keep them quite discrete. Too much professional writing goes back and forth, repeating ideas from different segments. It makes it hard to follow, repetitive and sluggish. Be really disciplined about this.

4 Use a bit of word glue. In other words, when you're happy with the arrangement, make sure they link together seamlessly and feel like a flowing piece of copy. But don't use too much glue or you'll slow the story down. There are some examples of words

that link sentences together in the 'Make it flow' section of Chapter 13, Quick wins.

5 Segmenting your writing is a great opportunity to do some valuable pruning. Instead of just cutting words, could you cut whole segments? What's the least important segment in your piece? Could you lose it altogether, to give the rest more attention and prominence? Does every segment really have a strong role to play, or are some waffle or repetitious or tangents?

6 Your structure 'shouldn't show'. Structure is the spine that holds your business writing together. And you can't see your spine, you just benefit from its invisible support. Aim for a seamless, hidden structure that does three things:
- makes it easier for your reader to follow your story;
- arranges your segments to achieve maximum impact;
- ensures that your proposition is a constant thread throughout.

7 Make sure your best bits aren't buried. You're fighting a battle to capture your reader's approval – no point in waiting 'til late on, when the fight might already be lost. Look at where your best segments are – the big guns that are going to blow your audience away. Make sure they're near the front. Or, if they're late on, you should be utterly confident that the preceding segments draw the reader in so brilliantly, those big guns will still hit their target.

8 Fresh eye. Structure in particular is one of those areas which can benefit from having someone else read your writing. 'Can't see the wood for the trees' is an apt expression here: you can get so close to tweaking an individual sentence into perfection that you can lose sight of the fact that a quick bit of structural surgery could make a bigger difference.

And that's structure. Finding the best structural arrangement for your communication will really lift it, but sometimes people get very fixed ideas about what order you should tell the story in. Be open-minded to the idea that a communication's structure can be as individual as the piece itself – and at this *doing* stage, a little experimentation can pay dividends.

Essential example

The film *Gandhi* has an unusual yet effective structure. It begins with him getting shot. Then goes back in time 55 years, and the rest of the story is in chronological order, building up to repeating the scene where he gets shot.

9

How to craft that draft

So, thanks to your structural machinations, you've got a first draft typed up, with the segments in the order you feel works best. Now it's time to craft those segments a little. There are seven such areas of content craft I'm going to cover with you here. *Headlines, openings, furniture, clarification, objections, alternate view and closing.*

Headlines

Since your headline is the first thing your audience reads, it's one of the most important lines you'll write. No pressure.

> **Essential tip**
>
> Your headline is probably the first thing your audience will read – but it doesn't necessarily need to be the first thing you write. You might begin with a 'placeholder' headline but, as you write the main content, get an idea for a much stronger headline to replace it with.

So what makes a good headline? Well, there are three principles to crafting a good headline:

1 *The right length.* Yes, I know that sounds rather vague. But a headline shouldn't be so short it doesn't tell you anything, nor so long that it's more of an opening paragraph than a headline. Typically, look for anything between five to twelve words.

Doesn't have to be of course – the famous Wonderbra ad was just two words, 'Hello boys'. Whereas another famous ad headline from yesteryear is the fifteen-word classic, 'They laughed when I sat down at the piano. But when I started to play! –'. (Also reminds me of the Bob Monkhouse joke, a great example of wordplay: 'They laughed when I said I wanted to be a comedian. Well, they're not laughing now.')

Essential tip

If your headline is very short then you might take the opportunity to add a couple of words that add impact. However, if your headline is rather long but you can't find a way of taking out any of the substance, try rephrasing it, perhaps like a newspaper headline or Tweet would, to save words.

2 *Interesting.* Think of your headline and any subheads as adverts for the sentences and paragraphs which follow them. Adverts have to grab attention and interest their audience – so should your headline.

A simple way of getting a good headline is to look through the text that will follow it and find the most interesting thing within it. Then look at how you can say that interesting thing in the headline, only shorter (since the main text then expands on it) and as dramatically as possible. A dull, matter-of-fact or 'template' heading (like 'Introduction' or 'Summary') is like advertising that the text which follows isn't worth reading. In the headline, give yourself a bit of licence to be as potent, pithy and provocative as you can.

Essential tip

Once you've written your headline, try rewriting it different ways to find the strongest iteration. For instance, your headline might be 'A report into why our market share continues to slide'. There's no need to label that the document is 'a report'. After all, you don't start a conversation by saying 'I'm speaking to you today . . . ' It's also vague: 'sliding' doesn't tell you by how much. And it's all problem, whereas actually you'd expect the report to have recommendations for turning things around. So 'A report into why our market share continues to slide' could become 'HALT: How we'll stop our market share sliding within 3 months'.

3 *Relevant*: Your headline should always be relevant to the subject and the audience.

Essential example

I saw an ad the other day for home insurance. It had a picture of Scottish moorland, with the headline 'Home insurance that's a breath of fresh air'.

You may laugh (or not), but I've seen plenty of work like that. OK, the headline tells you that it's about home insurance, but 'a breath of fresh air'? What does that mean? What relevance has it really got to the message? It was just contrived so they could show a picture of a landscape, because they couldn't think of anything to do with home insurance.

It's like those pieces that have the word 'refreshing' in the headline so they can show a picture of a cup of tea. Or 'We're hammering down the cost' and showing a picture of a hammer. Even though the piece is actually about loans.

Personally I don't find those lines the least bit interesting, but I think they were created because the writer was trying to be *interesting* and did so at the expense of being *relevant*. A good headline will be both interesting *and* relevant.

As well as these three basic tenets, there are some headline categories I can suggest to you. Many of those categories correspond closely with the concept types we looked at in Chapter 5 of the *Planning* section of this book.

For instance, there's the *benefit* headline – a headline which promises the reader a benefit. As with your concept, find a way to dramatise that benefit as powerfully as you can. For instance, 'Helping you win more customers' is a benefit-led headline, but 'We've got 500,000 loyal new customers for you' is more dramatic. By losing 'helping' it becomes more of a promise, by being specific with the number it's more intriguing, and by adding 'loyal' we've included another benefit.

Other categories of headline include:

- making a guarantee: 'Twice the power … or we'll give you twice your money back';
- highlighting a problem: 'Losing money because your database is slow and unwieldy?';

- offering a solution: 'Save around £20,000 per client with our new datamart software';
- telling a story: 'The Soldier, the Sailor and the Supermarket Chief';
- having a powerful quote: '"We don't do market research" – Steve Jobs';
- using a surprising fact or statistic: '68 per cent of your customers can't even remember your name';
- asking a question: 'Is Google making us stupid?';
- posing a challenge: 'Have you got the courage our company needs?';
- intrigue: 'The dragon which played with fire';
- being provocative or shocking: 'Our latest vacuum cleaners really suck';
- apologising: 'We're sorry for what's about to happen';
- How/Why/When/What: 'Why everything you know about the soft drinks market is a lie'.

That doesn't cover every kind of headline of course, but it should be a useful start. And it depends on the context – some headlines can be quite plain yet effective. I saw one recently on an ad that just said 'Drink problem?' before offering an alcohol advice helpline number. What made it so powerful was the ad's location – stuck on the side of a bottle bank.

Openings

In an ad your headline has to be interesting to get someone to read on. But in a business report, where you're writing something the audience may have asked for or be expecting to see, they'll read the opening even if the headline is only so-so.

But, and here's the critical thing, if your opening is dull, they'll quickly start skimming. Which means, for some business writing, the opening can be even more important than the headline. So spend time really crafting your first few paragraphs – spend a disproportionate amount of time getting them right. And when you've written the whole piece, go back to the opening and see if you're still happy with it.

You're juggling a lot of different elements in powerful business writing – proposition, audience, tone, concept, the strengths of the medium and so on. For the opening, I suggest you just concentrate on building upon the headline and the concept, thinking all the time *what is going to get my reader absolutely hooked?*

Essential tip

I'm sure you've heard the term 'killer app'. The idea that something (originally technology) has such a great feature, you'll buy the thing just for that one feature. For example, a new console game that's so good, people buy the gaming console just to play it.

Well, when you're looking back over your communication's segments, which one is the 'killer app'? What's the most jaw-dropping, traffic-stopping, eye-popping thing you have to say? Because maybe, just maybe, it'd make for a good opening.

We're not looking for dispassionate, slow-paced introductions. We're not looking for reflecting back what they already know or have asked for. Ignore any preconceptions you might have about a logical, chronological flow beginning with the premise or the background. Does 'background' sound like the segment that's going to get your audience hooked? No. Decide what is, and open with that – or at least allude to it.

Furniture

By furniture, I mean the stuff around your main text. Subheads, graphs, charts and other visual aids.

In *Planning* we looked at some basics of layout, but when you're crafting your communication there are a few more things to watch for.

If you're using subheads, try and use them in a consistent way. A newspaper might have one-word subheads throughout an article, for instance. Also try and keep your subheads reasonably regularly spaced. There's also the argument that subheads should work chronologically – that someone could almost read just the subheads and get a summary of the whole piece, in order.

For visual aids such as charts, graphs and tables, I have three simple tips. First, make sure they're correctly labelled – in the text and with the chart. It's very common to find that, after changing the structure, deleting sections, adding new ones last minute, the charts aren't correctly labelled. The text refers to 'Fig 6' but there is no Fig 6, just a Fig 7.

Secondly, don't show them at a size or level of detail that makes them impossible to read. I've specifically mentioned the PowerPoint example before, where a presenter says, 'Now, you won't be able to read this, but ... '. The same goes for any business communication – proposal, report, briefing, whatever; don't show charts that have a little of detail and scale that doesn't work at the size they'll appear.

And lastly, give your visual aids meaningful, explanatory titles that sit with them. Don't assume I'm as clever as you, able to glance at a table and work out that a quarter more people are buying new product X, but that it's cannibalised sales of Y which is actually a more profitable line. Instead, tell me that in a caption next to the table. *Table showing that sales of X have cannibalised sales of Y, depressing our profits by 3% in the last quarter.*

Think about the other bits of 'furniture' too, from covers to page numbering. Many good examples of business writing are let down by oversights in these details – because they're not part of your actual business writing, they get overlooked; quickly thrown together at the last second. Yet they can often be very visible and they impair the impression of your whole communication if they've been neglected.

Essential example

Recently I was involved with a very high-profile report where the client insisted on having 'Contents' listed within the contents. In other words, on the page 3, the heading was *Contents* and the first listing under it was *Contents: page 3*. Crazy.

Clarification

OK, I'm back to the recurring theme of simplicity. When we looked at usage, I promoted the idea of using simple language, short words, shorter sentences and simplifying ideas. But there's more to consider when you're crafting powerful business writing.

It's about looking at every segment and asking yourself *am I saying what I mean as clearly as I possibly can?* Jargon, obfuscation and passive voice aside, much business writing still contrives to explain things in as clumsy and confusing a way as possible. Let me give you a real-life example.

Essential example

I live in a new-build estate. Here's the letter the site manager put through my letterbox (and everyone else's). I promise you, other than not reprinting the phone number, I haven't changed a single word:

> *We are going to be undertaking the paved surface works on your development week commencing 21st March. We anticipate this will take approx 8–10 days.*
>
> *It is appreciated that these works may cause some inconvenience to the residents and I apologies in advance should this be the case but in order for us to carry out the works effectively can I ask that you refrain from parking any cars on the road throughout the working day.*
>
> *In the event of any questions of concerns, please contact the Customer Care Line on XXXX XXXXXXX.*

It's kind of impressive how many mistakes the site manager has managed to shoe-horn into three short paragraphs. Using 'apologies' instead of 'apologise' and 'questions of concerns' are the least of its problems. Here are just a few of its other shortcomings:

'We are going to be undertaking the paved surface works ...' This doubling up of verbs ('going' and 'undertaking') happens a lot in business writing. 'I'm hoping to be able to show you' for instance – 'hoping' and 'show'. Just say 'I'd like to show you ...' Or 'X works by looking at ...' Just say 'X looks at ...'.

Here, he could have simply said 'We will be undertaking'. Of course, 'undertaking' isn't a great verb. It would have been better to simply begin 'We will be starting the paved surface works on 21st March'. It would have also saved him adding yet another verb, 'commencing', at the end of the sentence.

'Paved surface works on your development' – do you know what that means? It means they're going to block-pave the road outside my house. So why not say that? Clearer, so I'll understand it more easily. This goes on throughout, using words like 'refrain from'. It's just trying to be polite, or perhaps pseudo-posh, but there's really no need. Instead of 'refrain from', just ask me not to park my car on the road. In fact, since that's a pretty important action you need me to take, perhaps it should be in a headline.

The second paragraph is a single sentence, 51 words long. That's because it contains four separate pieces of information – that they realise it will cause inconvenience, that they're sorry about it, that they want to carry out the work effectively and that therefore they need me to not park my car on the road.

'It is appreciated,' is of course a passive style – losing the 'I', 'We' and 'You'. As is 'the residents'. He doesn't mean residents, he means me – so again, for clarity, say so. That whole section isn't really needed either – why not simply begin the paragraph at 'I'm sorry for the inconvenience, but ... '. In fact, a useful headline might have been something like '*An apology ... and a request*'. That would have got me reading on.

You see what I mean? It's just three paragraphs, about something as simple as paving my road. Yet ten minutes of editing would have increased its clarity – and therefore its effectiveness – tenfold.

Of course, there are occasions when you might, possibly, maybe, perhaps, shun clarification. I read this newspaper report a while ago: '"My understanding is he asked for the ordnance to be delivered between where he was located and where he saw people coming towards him," the officer said.'

Do you know what 'delivering ordnance' means? It's military speak for *dropping bombs*. Because the story was about 'collateral damage' – military speak for *we hit some civilians by mistake*.

Answer objections

Essential example

'Yes, but what about ... ?' 'I don't understand why ... ' 'Isn't that going to be too expensive/time-consuming/difficult?' 'Really? Can you prove it?' 'Shouldn't we X first?' 'What does Y think about this?' 'Isn't Z a better choice?'

That list could go on forever – it depends on what you're writing. But the principle remains the same: your audience will have doubts. Questions. Objections. In fact, I'm sure you've been on the receiving end of a business communication that you had some questions about, or that you saw a flaw in that wasn't addressed. And perhaps you later had a meeting with the author and you said, 'Hey, but what about

X?' and they gave you a perfectly good explanation that allayed your concerns.

Wouldn't it have been better if they'd have allayed your concerns *in the communication*, before you'd decided that it wasn't right?

So when you're crafting your work, consider what questions and objections your audience might have. Prepare your answer, and include it in your piece. You can be explicit about it, with something like 'You might be wondering how ... ' or be a little subtler. The point is, if you're making a case for something, write out a list of the audience's most likely objections or disagreements with it and craft your words to counter those objections right there and then. Perhaps with a line about getting in touch if they want to know more.

Alternate view

We looked at the audience you're writing for in *Planning*. But now we're *Doing*, I want to remind you of something about your audience: you're not them.

If you're a senior manager in a company, writing a piece for other senior managers at the same company, then from a purely 'pen portrait' point of view your audience are a lot like you. From that perspective, you would be writing to yourself.

But you may well have done some personality profiling at some point in your career – Belbin or Myers Briggs or similar. In which case you'll know what kind of business person you are – whether you're all about the big picture or the detail. Whether you're good at starting things but not finishing them. Whether you're action-oriented or thought-oriented. And whether you seek breadth of knowledge or depth of knowledge.

As you craft your communication, bear in mind these personal preferences you have. Remember that your audience may well be a different personality type. And that to win them over, you'll need to appeal to their personality as well as your own. If you have a tendency to cover a depth of knowledge, pull back a little and include some breadth too. If you're very action-focused, step back from pushing for things to be done too quickly and inconvertibly for those people who prefer to think things over for longer.

It's not something you can worry about too much – you'd be paralysed with indecision, trying to decide how to phrase everything.

Just bear it in mind, and remember to include a nod to audiences who come at things from a different perspective to you.

Closing

I'm sure you've heard this presentation mantra: 'Tell them what you're going to tell them. Tell them. Then tell them what you've told them.'

It's a useful *aide-mémoire* for powerful business writing too. When it comes to finishing your communication – whether it's an ad, a B2B (business to business) proposal or a staff memo – remind people what you've told them at the end.

Essential example

Say you were writing a B2B piece to persuade another company to partner with yours. The benefit to them is that they'll be able to enter a new emerging market for about £200,000 less than it would have cost them if they'd gone it alone, and they'll also have a greater chance of success. You've got all that in the main text – in fact you led with those powerful benefits up front. Now at the end, remind them of these benefits – to enter an important emerging market, to have greater surety of success and to save two hundred grand into the bargain.

You might use this reprise to phrase the benefits or the opportunity in a slightly different way than you have elsewhere – partly to sound less repetitious, and partly because as I've already said, different people process information differently, so it's wise to mix up the ways you present your case.

And of course, have a strong call to action in the close. Make it crystal clear what you want the audience to do, and ally taking that action with achieving the benefits you've laid out in your document. Give your call to action some crafting: play with ways of phrasing it that make it as natural as possible. Associate it with positive emotion and with perfect logic. And tinker with it so that it sounds neither too strident and demanding, nor too wishy-washy and unimportant.

And those are seven areas to consider when crafting your draft.

Essential tip

Here's an extra idea for you – productize. That may not be a real word, but I'll use it anyway. It's something I've found useful on occasion – and that's to give an idea a snappy name. When, in 2010, David Cameron became Prime Minister, he started bandying around the phrase 'Big Society'. In truth, few people knew what he meant – and he didn't go to great lengths to explain it. But it sounded good at the time. It sounded like the kind of thing the country needed.

Or how about this: creative agencies sometimes offer 'the general public' a prize for coming up with an ad idea. They call it 'crowd sourcing'. Gives it a 'product name' that makes it seem sexier than it really is.

I've argued against jargon in your writing, but if you can introduce a pithy new term or acronym, clearly explained, it can be really useful. Catchy names are called that for a reason – they catch on and gain a kind of cachet of their own. 'Slacklining' for instance: do you know what that is? It's walking across a low, elastic tightrope. It's become popular and cool, and I think the cool name helps.

So, consider the occasional attempt to 'productize' an idea. It worked for Twitter. They called writing a message 'tweeting'. Making it sound distinct from 'instant messaging' or any other generic product category term.

10

How to win hearts & minds

WIIFM, Maslow, universal motivations

A quick question before we continue. What's the point of your business writing? To inform? To entertain? To make an announcement? Because we've got some marketing budget to spend?

We may be doing a little of all of those, but only in order to achieve the real goal: to *persuade*. Right at the start of the book, I described powerful business writing as telling *a simple, engaging, persuasive story*.

Your communication must persuade someone to do something that otherwise they wouldn't have done. Your writing must change behaviour. Your words must *sell*. That's not easy – which is perhaps why some people forget that's the real purpose of good business writing. They settle for communications that are merely informative or entertaining or brand-building or awareness-raising and hope that their audience will just 'buy'.

But they won't. Not unless we sell. And an absolutely key principle to effective selling – whether you're selling a proposal, an idea, a recommendation, an investment, a product or a service – is to win your readers' hearts and minds.

In fact, we sometimes talk of 'head, heart and hand': win their head and their heart and you'll get them to put their hands in their pockets and buy what you're selling (or do whatever else it is you want them to do).

Or, as former Brazilian President Lula Da Silva put it, 'Pockets are the most sensitive parts of a human being. So we must touch hearts and minds first'.

Essential example

It's said that people buy cars for emotional reasons, but justify them for rational ones. Something that people who make car ads know all too well.

When you buy a new car, you'll actually be driving it in rush hour on the M6 in pouring rain and realising that the windscreen wipers are a bit ropey. But, in the ad, they show you the car winding through a glorious mountain pass in the South of France, with not a single other car in sight.

And they tell you almost nothing about the specifics of the car – even though it's likely to be the second most expensive purchase you'll ever make. They just focus on associating the car with positive emotions: joy, freedom, excitement, sense of achievement, feeling special. And it works. It drives you (pardon the pun) to the showroom, where to accompany the emotional reasons that got you there, you'll get a brochure full of rational reasons.

So you can buy the car because you enjoy how it makes you feel – but you've got a whole bunch of facts and stats to rationalise the purchase (to yourself and others).

In fact, it's often been found that emotional reasons to do something are more persuasive than rational ones. Let me say that again, because it can be something of a revelation: *emotional persuaders are usually more potent than rational ones.*

The reason it's a revelation is that in our modern high-tech world of science and research and sophisticated media-savvy consumers, we've come to believe that we're rational creatures. That logic is best and reason always wins. And everyone likes to think they make decisions logically, not emotionally.

But it's not true. In fact, studies show that our brains process information emotionally first and rationally second. It's an evolution thing: fight or flight. You need to know in a heartbeat whether to flood the body with adrenaline and fight the T-Rex or run from it. (OK, I know there weren't any T-Rex's in human evolution. All the dragons had killed them.)

Which means that a) we can't help but process information emotionally, and b) emotions are powerful influencers. Put the two together and you can see why it makes sense to use emotion in every communication – even though your instincts may be to write purely logically and rationally in business.

To make this work, simply list three things:

1. The biggest rational reasons for the reader to say 'yes' to your communication (eg, 'because my recommendation will save the company 20 per cent on manufacturing costs within 12 months');
2. The biggest emotional reasons for the reader to say 'yes' to your communication (eg, 'because publicly they'll get the credit for lowering costs and being able to avoid making anyone within manufacturing redundant');
3. The strongest emotions you could see working in your communications (eg, 'joy', 'fear', 'anger', 'excitement').

Then weave those into your communication – it really makes a big difference to consciously decide which emotional hooks you're going to use in your writing.

Essential tip

Don't *say* the emotions, make your audience *feel* them. Business writing can be littered with phrases such as 'we are passionate about', or 'this is a very exciting time for us', and sometimes even a smattering of exclamation marks to convey energy. The result is unconvincing and mawkish. Think 'How can I make this idea seem exciting' without simply describing it as 'an exciting idea'.

But let me tell you a secret. There's one big winner when it comes to capturing hearts and minds. One easy, fail safe way to shape the underlying theme of your communication and help you decide which emotional and rational levers to pull.

And that's to answer the question 'What's In It For Me?' (WIIFM).

It works like this. Just imagine (since it's likely to be true) that when your reader glances in the direction of your communication, that's what they'll be thinking. 'What's in it for me?'

You've got to answer that question. At the beginning, in the middle and at the end. And everywhere in between. And what's in it for them is what will benefit them, emotionally and rationally, from reading your communication and taking the action you want them to take. Here are two powerful ways of answering an audience's cry of WIIFM.

Maslow

As you may well know, Maslow was a psychologist who, in 1943, categorised the basic needs of humans and put them in an order. With the most fundamental needs at the bottom of a pyramid, and the most esoteric needs further up. He called it a *Hierarchy of Needs*.

The idea is that only when people have satisfied their needs on one level will they start worrying about the needs on the next level up. There are lots of versions of his pyramid, but most look like this:

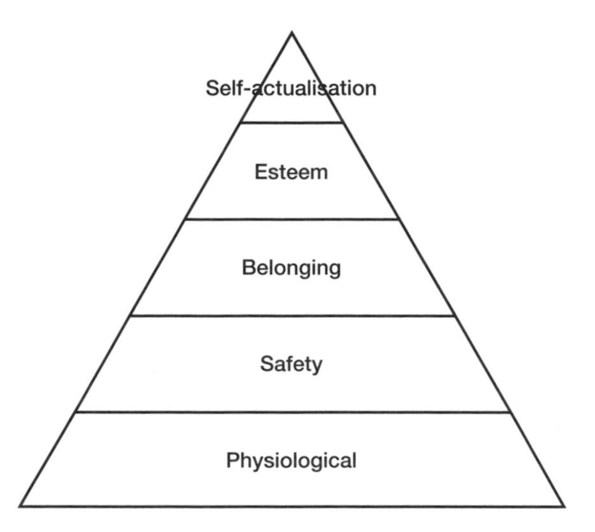

Why is this important? Well, it's important to know (and tell your audience) which of these needs will be met by your product/service/offer/recommendation/proposal.

It's also useful because sometimes you may be helping your audience meet more than one need on the pyramid – in which case, lead with the need that's lower down. Gather your content around that need, because it's more fundamental and therefore more persuasive.

Essential example

Say you were doing something for a charity, Oxfam maybe. An appeal to help the people of Sudan, who are in desperate need of clean, safe water.

Clean water is right at the base of Maslow's pyramid. People know it's a fundamental need. It's likely, therefore, to be a more powerful, persuasive story than talking about helping people in Sudan who've been separated from their family to find them again. That's from the 'love/belonging' level – higher up and therefore less of a priority. So focus your communication on the lack of water, not the family separation.

So the Hierarchy of Needs can give you a steer on the most persuasive content when talking about the recipient of the reader's charity. But it can also directly help answer the reader's question: 'What's in it for me?'.

Which might surprise you, since in a charity appeal you're asking people to make a donation for which they (in theory) get nothing back. It's not *their* physiological needs which are being met after all, it's someone else's – someone they don't even know. What *is* in it for the reader is a boost to their self-esteem.

That's why charity appeals should make it clear how much the audience is appreciated, how wonderful they are. The audience can meet a self-actualisation need (morality and personal fulfilment), if we tell them just what a fantastic difference their support is going to make (and clearly show them how).

And that's really useful, because it means you can get both a physiological need for the people of Sudan *and* a self-actualisation need for the audience into the story.

Now. Maslow's theory has been around for a long time and is widely used, but it's not always helpful. For instance, if you were doing a TV ad for McDonalds, where does that sit on Maslow? You're advertising food – does that mean you're meeting a physiological need? Well, fast food is eaten for convenience and pleasure, not because you'll starve otherwise.

So actually, McDonalds' advertising, in Maslow terms, might be more about belonging or self-esteem. But writing an ad for a Big Mac around those themes might not lead to very effective work. Luckily, there's another way to look at WIIFM.

Motivations

If Maslow is about *needs*, we could also create a list of universal human *wants*. It might include, in no particular order, these 32 motivations:

1. to be liked
2. to be loved
3. to be popular
4. to be appreciated/valued
5. to be right
6. to feel important
7. to make money
8. to save money
9. to save time
10. to make work easier
11. to be secure
12. to be attractive
13. to be sexy
14. to be comfortable
15. to stand out
16. to fit in
17. to be happy
18. to have fun
19. to gain knowledge
20. to be healthy
21. to satisfy curiosity
22. for convenience
23. out of fear
24. out of greed
25. out of guilt
26. to belong
27. to gain respect
28. to avoid pain
29. to get pleasure
30. to give life meaning
31. to achieve something
32. to win

What content can you pull together to promise people one or more of these things? How will reading what you've written help them achieve something, for example, or make them more popular?

Back to the McDonald's TV ad. Now that we're not looking at a Maslow physiological need, we might look at number 29 from the list of motivations: pleasure (the taste) as the thing we will promote in our ad. Or even number 17: because it makes you happy. Don't underestimate this stuff. It's incredibly powerful because it's what people – your audience – *want*.

The list of motivations works in every medium for every audience and regardless of whatever trends are happening. Why? Because these are things that have motivated humans for thousands of years. We might believe we're all very sophisticated and media-savvy, but the basic blood-and-guts motivators are in our DNA, in our evolution, and they haven't changed.

It's why those spam emails continue to draw people in.

The Nigerian princess who promises you millions of dollars if you give her your bank account details. The get-rich-quick business ventures which claim 'I'm making $15,000 a week online, just by working a few hours doing something I enjoy!' The Viagra emails that say – and I quote – 'D'UWantTtoHavePpeferctSexAllNightIonng?'.

They all play on promising people something they want. A motivation they want to fulfil so badly, they're prepared to ignore their better judgement. Even though it's telling them that if an offer sounds too good to be true, it probably is.

11

How to be utterly fascinating

Seven fascinators that will keep your reader hooked

'You can't bore someone into buying your product.' You might not be writing literally to get someone to buy something, but the premise holds true: powerful business writing is powerful partly because it's worth reading. It's interesting. Better than that: it's gripping. Compelling. Fascinating.

As we've already explored, the best way to be interesting is to talk about something that interests your audience – such as 'what's in it for them'. *What* you say is the most important thing, as we explored when we looked at propositions in our *planning*.

But after that, the *way* you say it can make a substantial difference – through your writing's personality and tone of voice, which we've already covered, and with seven proven fascinators we'll look at here.

They're not necessarily connected to structure or substance or personality. And they're not particularly based on principles of psychological persuasion. They're just useful devices you can write into your communication, to turn the humdrum into a humdinger.

Big bang

Start with a 'big bang'. Think of your communication as an action movie that starts with a huge explosion that makes the audience go 'woah'. That means an approach which doesn't build up to the subject: it hits the ground running with something as dramatic and attention-grabbing as possible.

Anecdotes

A moment ago I was talking about tone and I gave the example of the time I wrote as Lady Thatcher. An anecdote which, I hope, made explaining tone more interesting.

After-dinner speakers: most of their act is a collection of anecdotes. In fact, an anecdote lifts almost any presentation; it gives it colour and personality. And it's the same for your writing. Some anecdotes can be about your own experiences. Others can just be stories from anywhere – anything that helps illustrate your subject.

Essential example

I once wrote a piece that was about honesty. In it, I related an anecdote about Koko, the famous gorilla. What's she famous for? Well, Koko, a lowlands gorilla who lives at San Francisco zoo, is fluent in sign language. And she can understand around 2000 words of spoken English.

She also has a pet cat, given to her when it was just a kitten. One day, Koko's keepers heard a terrible commotion coming from Koko's enclosure. They went to investigate, and saw that the sink had been ripped off the wall in a tantrum.

They scolded Koko for her fit of pique. At which point Koko pointed at her tiny little kitten, and signed that no, Koko hadn't broken the sink . . . the kitten had done it.

So there you go; an anecdote about a fibbing gorilla that made a piece about honesty more interesting.

Water-cooler moment

This is the idea that people at work, all gathered around the water cooler first thing in the morning, talk about whatever's on their mind – whatever's grabbed their interest.

It may be last night's football game or episode of *EastEnders*. Or, if there's a big story in the news, it might be something to do with that story that they think the other people might not have heard. Something that might impress or interest them.

Because one of the reasons people like interesting snippets is so they can pass them on, and seem interesting by association. Which means facts can be interesting.

Essential example

Did you know that Wembley Stadium has more toilets than any other building in the world? (2618 apparently.)

Water-cooler moments can include:

- *Facts*: 'Seven reasons why our new 3D smartphone technology will change the market forever'.
- *Statistics*: 'Our new 3D smartphone technology is 82 per cent more powerful than anyone else's'.
- *Revelations*: 'Confidential: the first glimpse of our top secret 3D technology'.
- *Rumours*: 'Why those in the know are saying we're about to launch a bid for Nokia'.
- *News*: 'Our new 3D smartphone technology launches worldwide next month'.
- *Controversy*: 'From today, we're going to make Nokia's technology look like a kid's failed science project'.
- *Clever examples*: 'Now you can take a photo with your phone and have it forever in 3D'.
- *Clever solutions*: 'We've turned 2D into 3D by copying the way a bat uses echoes'.
- *Clever explanations*: 'Why our new 3D technology looks like the way a bat hears'.

Basically, giving your audience a new, bite-size, soundbite piece of knowledge they can either personally make use of, or pass on to someone else. It will always help make your communication more interesting and better appreciated.

Intrigue

I said that a 'big bang' opening is to liken your communication to an action movie.

You could also think of your piece as a thriller. One with twists and turns. With surprises. Where plot-threads come together over time, and where the picture builds gradually, as new pieces of information are revealed.

Good business writing can do a little of all of those. In other words, there's no need to put all your cards on the table right away.

So to keep your audience hooked, keep them on tenterhooks.

How? Well, not by padding the communication out or being deliberately confusing or misleading. Instead, consider the *tease*, where you hint at something important or surprising or a great solution, but don't reveal what it is straight away. I'll explain why in a moment.

That was a simple example, by the way, writing 'I'll explain why in a moment'. Another might be 'Before I explain how we're about to double our profitability, let me ask you a question'. Or 'But first, I want to ...'.

Telling the reader the number of items in your list works too. OK, it might not be fascinating, but saying 'Five tips that could extend your lifespan by as much as a year each' is more compelling than just saying 'How you could expand your lifespan by up to five years', because it lets the audience know there are five answers, and they'll want to read them all, just in case the fifth one turns out to be the best.

As well as the *tease*, there's the *twist*. Where you seem to be building up to one conclusion ... but then confound your audience by revealing the answer to be something different.

Essential example

Many jokes work this way – here's one of Jimmy Carr's: 'A lady with a clipboard stopped me in the street the other day. She said, "Can you spare a few minutes for cancer research?" I said, "All right, but we won't get much done".'

Thirdly, there's the *challenge*, where you challenge the received wisdom of the day – suggesting something that seems against what is already accepted or which is counter-intuitive. And of course, there's the *denouement* – when you make the 'reveal' and give your audience the big moment you've been building up to: a dramatic conclusion, a clear recommendation, a stunning solution.

And there's a well-worn phrase you'll already know: *leave your audience wanting more*.

Some business writing isn't specific and clear enough in its assertions – there are lots of vague truisms and unexplained conclusions. But even more business writing tells you too much – it goes into micro detail, and loses the focus on the big picture.

If you're a senior businessperson writing to other senior businesspeople, it's the big idea you should be discussing. With just enough detail to make sense of that idea and make it 'buyable'. But the chances are, you're not asking them to go and act there and then, the moment they've finished reading. You'll want them to call you, come see you, discuss it more. You want to intrigue them enough so that you can 'seal the deal' in person.

So be wary of telling them too much. Tell them what they need to know, what will build your case – but don't be exhaustive in your explanation. Even simple devices like writing 'I'll be happy to give you three really good examples at the next board meeting' will help create the intrigue that will 'leave your audience wanting more'.

Wordplay

As with personality, wordplay is one of those things that needs experience to do well – basically because it's easy to use it too much and too clumsily. It needs a light touch, to complement what you're saying, rather than *become* what you're saying.

Sometimes people who think of themselves as creative or witty try and imbue their writing with wordplay – and it becomes overpowering and irritating. Which, since you're trying to win your audience over, is exactly the opposite of what you want.

However, a little bit of lyricism can work well. In Chapter 13, 'Quick wins', you'll find 20 different types of wordplay to make your business writing more interesting.

Back to the start

A simple device you see commonly used in advertising communications and magazine interviews: the writing begins with a certain concept before moving onto the main topic.

Then, at the end of the piece, the opening gambit is reprised. It 'closes the circle' and just feels 'neat' for the reader. It ties the piece up in a nice creative bow, and makes it a little more interesting.

Essential example

A blog I wrote began 'When my mum was a girl, not many young people had watches. They were a bit of a status symbol. So in the summer, she and her friends would stick a watch paper template around their wrists. Which would give them a watch tan line – so it looked like you had a watch, you just didn't happen to be wearing it.'

Then it moved on to the main topic (about the difference between motivation and behaviour) without any further mention of watches until the very end: 'It's why understanding motivation, not just measuring behaviour, can be so important. And why people no longer buy watches to tell the time. But to tell you something about them.'

Quotable

Quotes always add interest, and they don't need to be from someone the reader has heard of.

Simply because they're speech they're direct from a real person, not a hidden writer, so they add gravitas to your words. They also add a different tone of voice to that of the rest of the piece. They can endorse what the rest of the communication is saying (effectively, you're writing something that's just agreeing with itself, yet somehow that works).

And it also means you can get away with saying things that otherwise you couldn't say.

Essential example

I've seen a skincare TV ad featuring a minor celebrity. At the end of the ad, she says, 'For me, it's the best moisturiser there is'. Of course, the company couldn't say it was the best moisturiser there is, that would be untrue or unprovable. But the celeb can say that, 'in their opinion', it is the best. And this ringing endorsement makes the whole story a little more compelling.

12

How to be irresistibly persuasive

Twelve psychological triggers to influence and compel your reader

Right: time for you to do voodoo. To climb inside your audience's head. And mind. And brain.

There are a number of psychological principles that really will enhance the persuasive power of your writing. Aside from any useful stuff I learned while doing a degree in psychology at uni, it also gave me a lifelong interest in the subject.

I've continued to hoard snippets of insight into human psychology ever since. What makes us tick and how we process information. It can all be used in business writing to make your story more compelling.

The result is a 'Dirty Dozen' of psychological triggers I'd like to share with you here. Well, they'd be a dirty dozen if you used your new powers for evil. You'll use them for good of course. Making them more of a 'Divine Dozen'.

They're all related to the way humans process information. There are very few absolutes in the human mind: most information is processed contextually. So when our business writing shapes the context, we influence the way someone processes the information contained within that writing.

Together, these twelve triggers are a fairly loose mix of cognitive psychology, developmental psychology, social psychology, pop psychology and pseudo-psychology that would have my lecturers turning in their corduroy-with-suede-elbow-patch jackets.

But it'll make your writing more hypnotic than Paul McKenna. Let the voodoo begin.

Popularity

It's this simple: people respond better to people they like than they do to people they don't like.

You'll do almost any favour for someone you really like. The office bore who also happens to have poor personal hygiene? Not so much.

Or imagine if a stranger tries to sell you raffle tickets. Again, you're less likely to say yes than if a friend offers them to you. The prizes haven't changed. The charity that the money raised goes to hasn't become any more worthwhile. But because someone you know and like has offered the tickets to you, you're much more likely to say yes.

The same is true here. Write in a way that makes people feel they know you and like you (or rather, your subject) and they're more likely to do what you ask them to do.

So think about how you might get someone to like you 'in real life'. Especially the things you say and do when you want someone who doesn't yet know you to like you.

Here are ten things you might do to become liked by someone, which you can easily replicate in business communications:

1. Use compliments/flattery (subtly).
2. Demonstrate shared interests/values/beliefs.
3. Show interest in them.
4. Be helpful (as in performing a chore or task for them).
5. Mention mutual acquaintances (they'll like you if you're liked by someone they like).
6. Use the same kind of language.
7. Be interesting.
8. Be successful.
9. Help them be successful (do something that makes them look good).
10. Establish familiarity.

All are pretty self-explanatory; it's up to you how you use them in your writing.

Essential example

One of the most famous opening lines in a direct mail pack was for American Express. It began, 'Quite frankly, the American Express Card is not for everyone. And not everyone who applies for Cardmembership is approved.' It's a great example of conveying exclusivity and of being subtly *flattering* by suggesting you might become part of this exclusive club.

Consistency

You often hear people complain about being pigeonholed. Yet the truth is, we like to pigeonhole ourselves. Because it gives us a sense of identity. Who we are and who we're not.

We like to think of ourselves as being such-and-such a person with such-and-such a personality.

And to help maintain that belief, we need to feel that we're consistent in our thoughts and actions.

So if someone has bought from you before, they're a customer – a customer who clearly values great quality. So remind them of that. Make them feel good about the type of person they are, and allow them to buy this new product from you, because it's clearly how they see themselves.

Brand 'fanboys' are an extreme example of this – people who see themselves as the type of person that the brand portrays their customers to be – and so will buy everything the brand produces, whether good, bad or indifferent.

Alternatively, if you can get some insight into how the audience see themselves (or would like to see themselves) then, again, you can write how your subject will help them continue to be that kind of person.

And if you know something they've done, you can use that to suggest that the action you now want them to take is consistent with what they've done before. Amazon does this in a pretty simple way, after you put an item into your shopping basket. The page reads, 'People who bought X also bought these items:'.

Essential example

'You're clearly someone who cares about our planet – and joining our campaign to stop rainforests being destroyed for palm oil crops was a wonderful way to make a difference. That's why I'm sure you'll want to help again today, by forwarding the attached message to the Chief Executive of EvilCorps, who are still destroying 100 hectares of rainforest a day to . . .'

Conformity

One of the things that defines us is which groups we're part of and which groups we're not.

People perceive us differently according to which groups we belong to, and because we know that, we aspire to be in certain groups (and aspire not to be thought of as part of other groups) pretty assiduously.

In social psychology, they call it 'Ingroup/outgroup behaviour'. The 'ingroup' is one you think you belong to (or aspire to belong to). The outgroup is one you don't belong to (or at least, don't want to belong to). People's actions can be shaped by wanting to be part of a certain ingroup or separate from a certain outgroup – and you can use that trigger in your writing quite easily.

Essential example

Let's look at the snippet about rainforests that I suggested for *consistency*, and change it slightly to use the *ingroup* trigger instead:

'Clearly, you're someone who cares about the destruction of the rainforest – and joining our campaign to stop rainforests being destroyed for palm oil crops was a wonderful way to make a difference. That's why I'm sure you'll want to join **the other people like you who took part in that campaign**, by forwarding the attached message to the Chief Executive of EvilCorps, who are still destroying 100 hectares of rainforest a day to . . .'

Of course, you can combine more than one psychological trigger – you could use both *consistency* and *ingroup conformity* triggers by writing: 'I'm sure you'll want to help again today, as so many others who took part in that campaign are, by ...'

Alternatively, the same basic copy could have used the *outgroup conformity* trigger instead.

Essential example

'Clearly, you're someone who cares about the destruction of the rainforest – and joining our campaign to stop rainforests being destroyed for palm oil crops was a wonderful way to make a difference. But today, we need to act again. **I'm sure you don't want to be one of those people who stands on the sidelines**, watching big business getting away with despicable acts of environmental rape. That's why I hope you'll forward the attached message to the Chief Executive of EvilCorps ...'

Brands often have a big element of ingroup/outgroup triggers to them. Are you a PC person or a Mac person? Marmite: you either love it or you hate it. And if you buy Armani jeans, you're unlikely to match them with a Matalan sweater. Brands show you an 'ingroup' and say 'If you want to be part of this group, buy our product'.

Essential tip

Look at the brand you're working with: what would its ingroup look like? Its outgroup? Then you can use the conformity trigger to add a little gentle persuasion.

Credence

If David Attenborough told me a population of orang-utans have been discovered in Indonesian Borneo that are evolving into a brand new species of human, I'd believe him. If my taxi driver said it, probably less so.

Credibility is very important in business writing – you're trying to persuade someone, so it helps enormously if they think a) your trustworthiness is credible and b) your authority is credible. Writing in a confident tone that shows you know what you're talking about and with the facts to back it up helps, obviously.

Essential tip

Somebody else endorsing your subject can help your communication's credibility – which is why testimonials work. You're putting words in the mouth of someone else like the audience, saying they (for instance) bought the product and they're delighted with it.

It's why we look at the reviews on Amazon, to see what other people have said about that *Greatest Eurovision Hits . . . Ever!* album we're thinking of ordering.

Using experts (like David Attenborough, above) works too. If you're writing a report to persuade the board to follow your recommended course of action, include the opinions of some other experts who agree with you. You could even use quotes from famous people where it helps. For instance, right at the start of this book, when talking about the importance of simplicity, I quoted Einstein: 'You do not really understand something unless you can explain it to your grandmother'.

Comparison

As I said when introducing these twelve triggers, there are very few absolutes in the human mind. We judge most things by comparing them with other things.

Which means you can influence what people think about something by determining what they compare it with.

Want to feel bad about your life? Think about the world's most popular, most successful movie star/film star, and what their life must be like compared to yours. Want to feel good about your life? Reflect for a moment on a child born HIV+ in a poor African village.

Rolls-Royce is a good example. If they have their £300,000 Silver Phantom at a car show, it looks a bit pricey compared to the others there. So what they often do is have them at boat shows. Where, if you're looking at a £4,000,000 yacht, picking up the world's most luxurious car seems like a snip.

And you'll sell more jeans costing £150 a pair if you also sell jeans costing £250 a pair – because by comparison, the £150 pairs will seem good value.

You can use our desire to compare things very effectively in your writing. If you're selling something expensive like a Roller, compare it with a yacht to make it seem relatively inexpensive.

If you're selling something cheap, like donating £3 a month to charity, compare it with the £3 they might currently spend on buying a glossy magazine. Or how much less £3 a month might be than how much they spend on take-away coffees. Suddenly just £3 to help save a life seems like nothing (and much more deserving, by comparison, than your skinny latte).

Compare the upside of having your product with the downside of not having it, to exaggerate its benefit. Compare where your product is close to another in performance, but cheaper. Or, if you're dealing with something expensive, compare it to something cheaper but not as good.

Essential example

Fairy washing-up liquid used to compare their product with the 'next best-selling brand' and demonstrate that It was better quality and so lasted 'up to 50% longer'. Which meant it was actually cheaper per use.

Compare how someone will feel by taking the course of action you want with feelings from some completely unrelated area.

Compare your new product with your old one: it's got 50 more horse-power, it lasts twice as long, it washes twice as clean, its graphics are 30 per cent more powerful. This comparison with your own old stuff is amazingly effective, because it says either 'Didn't buy before? You were right to wait ... the time to buy is now' or it says 'Bought our last product? It's out of date and you're behind the times ... get the new one'.

Compare before and after. Compare the benefits of your subject with similar benefits in a completely different area.

Compare to contrast. To exaggerate. To borrow brand equity or feelings or attributes from another category. Compare to highlight benefits. Compare to downplay disadvantages.

Loss aversion

Every year in the January sales, people in the UK spend over £600 million on clothes they will never, ever wear.

Bought, taken home, put in a cupboard and *never* taken out.

Why? Because no-one likes to miss out. And when you're at the sales, among the crowds, and people are jostling and elbows are flying and shelves are emptying, you get caught up in the moment. You get into a sales panic. 'If I don't grab something quick I'm going to miss out!'

So you buy something that, later, you don't even want.

Lots of 'Sale Ends Tuesday', 'Limited Time Offer', 'Only a Few Remaining' stuff works on that basis. And look at eBay: prices are often driven up by people getting into a frantic bidding war. It's not just that they want the item; they want to 'win' and not be the loser who misses out – even if that means they have to pay a premium to do so.

Essential example

Near the end of a piece I wrote for a credit card: 'As I'm sure you understand, we expect this offer to be very popular. For that reason, we're unable to guarantee how long we will be able to make it available, even to specially chosen customers like you. So, if you do think this opportunity is right for you, please apply as soon as you're able, to avoid missing out.'

Of course, your communication may not suit an 'everything must go' salesy tone of voice. In which case, you can still use this psychological trigger by conveying a little exclusivity.

Bargain

'It's a deal, it's a steal, it's the sale of the f**king century. In fact, f**k it – I think I'll keep it.' So says Tom in *Lock, Stock and Two Smoking Barrels*.

Getting a deal or a bargain is closely related to loss aversion. In the case of the January sales, for example, it's a bargain that you're afraid of missing out on. It doesn't have to involve loss aversion though; the feeling that you're getting a good deal is a powerful psychological trigger all of its own.

This doesn't mean you have to write in a way that suggests your subject is *cheap*; instead it can simply be *good value*.

When a pricey perfume (sorry, *parfum*) throws in a free bag, suddenly it seems like a bargain. When a charity tells you that you can save a child's sight for just 80p, that sounds like great value. And when Sainsbury's gives you a bunch of vouchers you feel you're getting a good deal; a feeling you wouldn't have got if the things had just been permanently at that lower price in the first place.

Once on a holiday at an all-inclusive resort in Mexico, I agreed to go to their 'sales presentation' (for membership to the hotel group to get big discounts on all-inclusive holidays). I just wanted to see their sales pitch. There were lots of selling techniques in there (such as asking you questions to involve you, asking questions you had to say 'yes' to and so on). But one was the use of a 'bargain' technique.

At the end of the pitch, the man told me the price. He said most guests were Americans, so he was used to giving the price in dollars, but as I was English, he'd give it to me in sterling. It was £15,000.

Once I'd stopped laughing, he said: OK, hang on a minute, how about this? What if he kept the number the same, but he made it dollars instead of pounds. $15,000 dollars – which was just over £9,000 – £6,000 less than it had been just moments earlier. Suddenly, it felt like a bargain!

Mutuality

We're an awfully polite bunch, us humans. If someone smiles at us, we smile back. And if someone does us a favour, we feel obliged to do them a favour back.

In fact, if someone does something nice for us, we feel in their debt ... even if we didn't ask or want the nice thing in the first place. Somebody adds a link to your blog, so you add a link to theirs. Somebody writes a great recommendation of you on LinkedIn and then asks you to recommend them ... you kind of have to, even if actually you didn't really rate them.

Essential tip

Give people things in your communication by telling them something interesting or useful. Entertain them. Make them smarter. Show them how to get a bargain.

Give them something for free and the mutuality gene kicks in: they'll be more inclined to do something for you. It might sound crazy, but humans are not rational beings. We're a bit mental when it comes to freebies.

For instance, have you ever had a charity mail pack with a pen in? Why do they do that? The pens are rubbish. Plus, you already have a pen. Several, in fact. So you don't need a rubbish one to fill out the donation form. And if you listen in on focus groups (as I have), you'll know that people who give to charities say time and time again 'Don't send me a pen – I don't want it and it's a waste of money'.

And yet, test two identical mail packs, one with a pen in and one without, and the one with the pen will get a better response. People feel indebted and give more. For being given a rubbish pen they didn't ask for and don't want.

Heuristic bias

Heuristics, in psychology, refers to the things that bias the way we make decisions. It's an amazing area of social science, and can be utilised in copy to make it more persuasive. One of those biases is called *anchoring*, and here's an example:

> *A group of people are asked to think of a random number and write it down. It can be any number they want. But, first, a numbered wheel is spun, and the result is given to the group. They're told this is just part of the process and that they are to completely ignore the number: it has no relevance whatsoever.*
>
> *Yet if the wheel was giving very high numbers, the group 'randomly' came up with a high number. If the wheel gave a low number, the group came up with lower numbers.*
>
> *They couldn't help but be influenced by the number, even though they knew it was randomly created and they were told to ignore it.*

For persuasive writing, it's a goldmine. Here are three types of heuristic bias you can use in your business communications.

Anchoring

As in the experiment with the number wheel, this is the tendency to allow one piece of information to disproportionately dominate our decision-making.

Ever worked somewhere where the new recruit sucked? You can see it, everyone else can see it – hell, even the new recruit can probably see it!

But you know who'll struggle to see it? The person who hired them. Because that bit of information, 'I hired this person', disproportionately dominates their judgment. They hired the recruit, so the recruit must be good. Because if they weren't good, then that would mean the person made a poor decision in hiring them. They can't admit that to themselves, so the fact that they did the hiring 'anchors' their thinking.

Availability

Information that stands out in some way also seems more available, and is therefore more influencing.

For instance, knowledge of our own driving is more available than knowledge of other people's driving. We're closer to our own driving. The result is that around 80 per cent of the population thinks their driving is 'above average'. Which of course it can't be. Since the average wouldn't then be the average.

In your communications, you can make information more 'available' by putting it in a headline or a call-out box or side panel or subhead; that makes it more influencing (it's why there are so many rules in financial advertising about talking about the disadvantages right alongside the advantages).

A second, subtler way is to use words that relate, in some way, to the subject, without necessarily saying them *about* the subject. OK, that sounds confusing – so try this on someone. Say to them, 'Spell the word "silk". Say the word "silk" three times. What do cows drink?'

People will often say 'milk'. Much more often than if you just ask them, 'What do cows drink?'.

You can influence your audience in a similar way by using language that fits with the things you want the audience to think about your subject, even when you're not talking directly *about* your subject.

Essential example

If you were writing about a product that you wanted to seem sexy, such as perfume, then use sexual, sensual language throughout, not simply to describe the product.

Maybe the ad is about a woman getting ready for a night out, ending with using the perfume, but before that they take 'a slow, steamy shower' and 'slip into a sheer, silken dress that clings to every curve' and they bite into a 'velvety smooth Belgian chocolate' before applying their most 'seductive, irresistible lip gloss'. Simply the *availability* of these words (the association, if you like) makes the perfume seem sexier.

A third way of using the availability heuristic is 'misleading vividness'. This is where making something more vivid in someone's mind makes it feel more likely than statistically it is.

For instance, ever told someone you were thinking of buying product X, and they said, 'Oh, I know someone who had one of those, it broke after 3 weeks! Terrible'.

That's pretty vivid available information, and it might well put you off buying the product. Because you've got some negative, first-hand available information on it. And you don't seek out the facts which show only 0.1 per cent of people have ever had their product X break. Unlike product Y which you end up buying, which has actually broken in 0.8 per cent of cases.

Essential example

'It could be you'. The slogan for the National Lottery makes the experience of winning seem very vivid, to the point where people talk about how'd they'd spend the money if they won. Yet ... well, look at London. It's pretty big and rather busy. Yet if *every single person in London* bought a lottery ticket with a unique set of numbers, there's still around a 50 per cent chance that *not one of them* would win the jackpot.

A fourth, very simple use of availability is the 'first and last' way our short-term memory works: when given a list of just about anything, people remember the first and last items better than any other positions. So in your writing make sure your most important points occupy the first and last positions in your list.

Risk aversion

Not quite the same as simple 'loss aversion', it's an important heuristic to know for good business writing.

It's the truth that in human minds, 'losses loom larger than gains'.

So, if you tell people they have an 80 per cent chance of surviving a surgery, they're more willing to undergo it than if you tell them they have a 20 per cent chance of dying.

Essentially you're telling them exactly the same fact. But in one you're talking about a gain, in the other, a loss. And even though the statistical probability is the same in both, the one where you refer to a potential loss makes it 'loom larger' in the patient's mind.

Essential example

Say you're writing a TV ad for home insurance. Instead of writing a voice-over that talks about gaining peace of mind, talk about how many people get burgled every year, how many house fires there are, how many houses suffer flood damage – all of which mean people can lose their most treasured possessions.

Of course, insurance doesn't stop you losing treasured possessions (it may be able to replace them), but this positioning of avoiding loss through the advertised home insurance will be very powerful.

Emotional decisions

You may not have had the pleasure of seeing the TV programme *Deal or no Deal*. Basically, it revolves around a load of shoeboxes that each have a different prize value written inside them. And the contestant has one of those boxes. The question is, does their box have one of the large prizes inside, or a small one?

Every round, they choose three of the boxes to open (not their own box). The values in them are revealed, gradually whittling down the possibilities for what is going to be in the contestant's box. And after every round, an unseen 'banker' makes them an offer for their box.

So, it's a game of random chance – there's no skill involved in deciding which boxes to open. The decision you make each round is whether to accept the banker's offer, or whether to play on until only your box is left.

The banker doesn't know what's in your box either, so he basically offers around the average amount of all the remaining values of the unopened boxes (sometimes slightly higher, sometimes slightly lower).

Say someone gets to the point where the banker offers them £46,000 – an exactly fair offer based on the values remaining. But the contestant decides to play on and open another three boxes. They open ones which turn out to be most of the high-value figures, meaning their box is more likely to have a low value. Damn, they've made a mistake. They should have accepted the £46,000 offer.

So now the banker makes a new offer: £21,000. Looking at the unopened values remaining, £21,000 is actually *higher* than the average. It's a generous offer. So the contestant should bite the banker's hand off and say yes. Walking away £21,000 richer just for having been on a gameshow for a day.

But they don't. The contestant isn't thinking about the 'generous' £21,000 offer. They're thinking about the £46,000 they were offered just moments earlier. They feel stupid. They feel angry. They don't want to accept £25,000 less than they were offered just a minute earlier. And they see that there's still one £50,000 value to be opened: maybe that will be in their box. Maybe they can still come out on top.

However, the odds are very much against them getting the £50,000. And in fact, after playing on for several rounds, they end up with just £200.

All because they had 'a rush of blood to the head'. They went from being rational and playing the odds to letting their heart rule their head and chasing the £46,000 prize they'd turned down.

It's why there are lots of references to emotive writing throughout this book: *because emotions are very powerful influencers of decision-making.* So, just as we covered in the chapter on 'winning hearts and minds', remember that people often make emotional decisions, even if they claim their decision was based on rational factors.

Essential example

A classic Nike ad showed a basketball legend leaping high, about to slam dunk the ball. The line: 'Michael Jordan 1, Isaac Newton 0'. Idea, image and line together stir the emotions – without having to use any emotive words.

Framing

This is the phenomenon that enables you to influence an audience's behaviour by the way you present the subject. Comparisons and leading with the potential loss rather than the equivalent gain are psychological triggers in their own right, but they're also examples of framing – presenting the information in a certain way.

Framing is an area that often needs you to think laterally and find a new way to present an option to make it more compelling.

For instance, have you seen the Volkswagen's Fun Theory 'Piano stairs' video on YouTube? On the underground in Sweden there's a set of stairs next to an escalator. And most people are using the escalator. So how do you re-frame the stairs, to make them a more attractive option?

They turned the stairs into a giant piano keyboard, complete with sounds when you stepped on them. And lo and behold, suddenly more people want to use the stairs. Giving them all a little extra exercise into the bargain.

In fact, what if that was your challenge: get people to take more exercise? You could spend lots of money telling them how important exercise is. Or you could just find ways to make things that cause you to exercise more appealing. Like using the stairs.

As I say, every challenge has its own unique framing opportunities, and you need to think how you might re-frame the action you want people to take in your communication – make it the more attractive option.

A smile in the mind

If you're happy, you're more likely to smile. But, weirdly, that wiring works the other way around too. If you smile, just the mechanical action of smiling (even without a reason) will often make you more happy.

It's this kind of feedback mechanism you can use in business writing to influence your audience. By giving them 'a smile in the mind' for whatever reason, you make them feel more predisposed to what you have to say.

Essential example

'"I never read *The Economist*" – Management trainee, aged 42.' 'Retire early with a good read.' 'What exactly is the benefit of the doubt?' Remember those classic *Economist* ads, type only (white on a red background to look like the magazine's masthead)?

▶

There were dozens of them and they were very witty. Sometimes they gave you 'a smile in the mind' because they were amusing. And sometimes they made you smile because they took a moment to work out, and when you did, you felt pleased with yourself for having done so.

Make your audience smile (in a way that's relevant) and you're halfway to having them sold. Not by telling them a joke (probably) but with a little wit, amusing insight or entertaining quote or anecdote.

As you can see, the psychology of persuasive writing is a vast subject – I only looked at a few types of heuristics, for instance, when you could write a whole book just on using those biases.

These 12 psychological triggers should be useful to you. But if you want to know more about psychology or how we can use the way our brains work to influence your audience then I recommend *The Decisive Moment* by Jonah Lehrer, *Emotional Intelligence* by Daniel Goleman, *Influence* by Robert Cialdini, *Nudge* by Robert Thaler and Cass R. Sunstein, *Blink* by Malcolm Gladwell, *The Happiness Hypothesis* by Jonathan Haidt and *Introduction to Social Psychology* by Miles Hewstone *et al.* (one of my lecturers from all those years ago at university).

13

Twenty-five quick wins

1. Start with a short one
2. Features tell, benefits sell
3. Avoid clichés
4. Be unusual
5. Metaphors, similes and analogies
6. Hardwired words
7. Wax lyrical
8. Solutions not problems
9. Nouns beat adjectives
10. Avoid talking about cost
11. Quantify
12. Be active not passive
13. Keep it short (or long)
14. Avoid a woolly ramble
15. Don't get them disagreeing
16. Don't know it, feel it
17. Show not tell
18. Three's the magic number
19. Tell them what you want
20. Urgency
21. The tease
22. Make it flow
23. Paint a picture
24. Reframe it
25. Back to the start

This is the little stuff; a grab bag of tricks and techniques that only take a moment to learn and which can improve your writing at a moment's notice. We've employed a little of their trickery already in this book (such as right at the start, with my 'flipchart exercise'), but here they are all brought together, for you to use as and when you see fit.

25 canny cantrips to add a final dash of potency to your prose

Start with a short one

Almost always.

Whatever the medium, I almost always begin with a short paragraph. Usually a short, single sentence.

One reason is that when someone glances at your communication, just a glance will be enough for their brain to process a short, stand-alone line. And if that first sentence is interesting, they'll start to be drawn into your story. However, if your opening gambit is a dense block of unbroken text, like an eight-line paragraph, then a glance won't be enough to take it in. And you've lost a small but significant opportunity to engage your reader.

In some ways, your first sentence can be like a second chance at a headline – a bite-sized morsel to whet their appetite and make them want to continue.

A second reason is that as well as *actually* being easier to take in, it *looks* easier to take in. It doesn't look like an off-putting wad of words. And it starts you off with the discipline of having some short, pithy paragraphs.

Like this one.

Features tell, benefits sell (& outcomes do well)

In the words of legendary advertising writer John Caples: 'Don't sell the world's best grass seed, sell the world's best lawn'.

The feature is a characteristic of a product/service/offer. The benefit is what that characteristic does for you. The outcome is what your world is like as a result of that benefit.

Businesses are closer to their own products and services than their customers are. As a result, as well as often talking in jargon or technical terms without realising it, they can become convinced of the importance of every feature of that product or service.

But to persuade someone to buy, we've got to tell them about the benefits or outcomes that will most appeal to them.

Just listing the features means you're asking the audience to work out what the benefit is for themselves – which they may not do.

And even if the benefit is obvious, it'll sound better when brought to life by you.

Essential example

'Our new handbag comes in four great colours' is a feature; 'Four great colours means you can choose the handbag that matches your outfit' is the benefit; 'You'll turn heads everywhere you go' is the outcome.

Avoid clichés

Avoid clichés and tired, overused phrases that are padding or have lost their meaning. As I've mentioned before, 'peace of mind' has become a cliché for describing the benefit of any kind of insurance.

'I'm writing to you today' is one you still see in letters. They know you're writing to them, so telling them something they already know is a poor opening. Using a cliché to do it is even worse.

> *As a matter of fact, before you know it, what you'll find is, fate worse than death, push the envelope, best-laid plans, it's not rocket science, with the best will in the world, more affordable than you think, there is a better way to ...*

These phrases lose their meaning. So the audience doesn't really take in what you're saying. It also makes your writing seem bland. If your writing is bland, your message is bland, so your product or service seems bland by association. So maybe I don't want it.

Be unusual

Flabbergasted. Cock-a-hoop. Blabbermouth. Obstreperous. Salubrious. High falutin'. Vituperative. Flim flam. Biggie. A huge fat wad of cash. A rare old time. A big beaming grin. A teensy weensy slice. Finger-lickin' good.

Words or phrases with a bit of personality, that stand out from the crowd, can make your communication more memorable. Even if people don't know the exact meaning of the word, they'll get the idea if it's in context.

Because, although you want your writing to be easy to read, you don't want it to slide straight off their eyes. You need a little friction for their brain to rub up against.

A moment's pause, courtesy of some striking language, will mean they're spending more time soaking up your ideas. Which is good. Novelists and poets often do it; describe something in an unusual (but evocative) way to really bring an idea to life. What we're looking for is the *opposite* of a cliché.

You can also use uncommonly used words to 'own' an idea – what company does 'exceedingly' make you think of? You can probably remember their whole strapline – not because it's a great line, not because the ads are great, not because their cakes are so delicious ... but because they used a distinctive word. (The same distinctive word, in every ad.)

If you want to be really distinctive, make the word up. Shakespeare did. He invented hundreds of words that we still use today. Including, appropriately, the word *articulate*.

Many ads are remembered for the catchy phrases they coined – the one I remember from my youth was 'Hello Tosh, got a Toshiba?' Although my prize for best made-up word sequence would go to Pepsi, back in 1973, for *Lipsmackin' thirstquenchin' acetastin' motivatin' goodbuzzin' cooltalkin' highwalkin' fastlivin' evergivin' coolfizzin' Pepsi.*

Metaphors, similes and analogies

As I said in *Planning*, metaphors and analogies can make good concepts, but not always. The problem is they require the audience to translate the metaphor into the thing you're actually talking about. Which audiences often won't do if they're not paying attention.

However, a metaphor (or simile or analogy) can definitely enliven your writing – in three ways.

First, the metaphor can bring an interesting new subject into your story. Secondly, the metaphor can be a clever comparison with the original subject which beguiles the audience. And thirdly, it can help the audience understand the point you're trying to make about the original subject.

Essential example

King of the metaphor/simile, Jeremy Clarkson on describing the (apparently unnecessary) suspension-adjustment lever on a Bentley: 'It's about as useful as putting a snooze button on a smoke alarm'. He could have just said, 'It's about as useful as a chocolate teapot', but that old simile has become a cliché – very tired and well past its use-by date. Instead, he came up with something clever and new that added life and meaning to his article.

Here's Charlie Brooker, being somewhat unkind about Cilla Black: 'She's starting to resemble the result of an unholy union between Ronald McDonald and a blow-dried guinea pig'. Or Howard Marks, describing himself: 'When I look in the mirror ... I see a not-too-recently excavated mummy of Mick Jagger'.

Of course, you can be metaphorical in a less explicit way.

Essential example

In the previous quick win, talking about unusual words/phrases, I wrote, 'Because, although you want your copy to be easy to read, you don't want it to slide straight off their eyes. You need a little friction for their brain to rub up against'.

Words aren't going to literally slide off their eyes, or literally rub up against their brain. It's a little metaphorical device, to illustrate the point.

Hardwired words

For some reason, some words are proven to get people's attention more effectively than others.

If you use one of those words in a headline, for example, there's more chance of it being read than if you don't. But remember 'relevant abruption'. Only grab attention in a way that's relevant. Some of the words that seem to be hardwired into our brains and command our attention include:

Free	*Enjoy*	*Save*	*Win*
Now	*New*	*News*	*Thank you*
Guaranteed	*Discount*	*Limited time*	*Last chance*
Introducing	*Urgent*	*You*	*Secret*
Promise	*Cash*	*Prize*	*Rich*
Sex(y)	*Breakthrough*	*Proven*	*Discover*

And of course, using your reader's name always has the power to grab their attention (one reason so many spam emails do it). Using the name of a current celebrity works too – that's why so many newspapers and magazines run 'non-stories' about celebs, because just using their name increases sales. See if you can discover more 'hardwired words'. (David Ogilvy claimed 'Darling' was one he'd found increased response when he added it to an ad's headline.)

Wax lyrical

Even if we have to say the same thing as everyone else, to the same people as everyone else, we don't have to say it in the same way as everyone else.

Take advantage of the richness of the English language to create rhythm, a little poetic flourish and the occasional moment of wordplay.

Essential example

An article I wrote used both sour and sweet in their non-literal sense together: 'And walking onstage to pick up an award while fellow creatives clap sourly is a particularly sweet moment'.

Why is a light touch of wordplay important, rather than a self-serving waste of time? Well, it makes your copy more interesting to read. More distinctive. And more memorable. And so, by association, the very subject you're writing about.

In other words, as long as you're being lyrical in a way that's appropriate for the brand's tone of voice, a little wordplay can be A Good Thing. Just, as I've mentioned before, be wary of falling in love with your own cleverness and indulging in wordplay that obscures or overwhelms rather than enhances. Less is more.

Here are 20 alphabetically arranged examples of ways you can make your writing more lyrically lovely:

1. *Alliteration* – words that begin with the same letter/sound. A pair of words together, as in 'dirty dozen' adds emphasis, but you can also use alliteration where the alliterating words don't appear together at all, but dispersed through a sentence, for a subtler effect. I saw one the other day: *Fantastic tasty Thai.* (They were referring to food. I think.) At first it looks like just

the last two words alliterate, but actually, it's five syllables in a row: tast-tic-tas-ty-thai.

2 *Antanagoge* – turning a negative into a positive. It can be really useful in business writing when you have to let the reader know about a disadvantage. For instance, 'Of course, it costs a little extra. But you'll save many times the price difference in just a year'.

3 *Antimetabole* – where you repeat the same phrase, but reverse the order, as in 'Ask not what your country can do for you, ask what you can do for your country.'

4 *Antiphrasis* – being ironic in a single word (like saying 'It's a *cheap* 10 million quid'). This kind of twist draws attention to the idea, to stick in the reader's mind.

5 *Antithesis* – an example is 'One small step for a man, one giant leap for mankind'. It just means to compare two ideas, to put them together to contrast them. Lots of aphorisms (wise sayings) are structured in this way.

   (Incidentally, as you may know, Neil Armstrong fluffed his lines – he said 'One small step *for man*, one giant leap for mankind', missing out the indefinite article 'a'. So what he said was actually a contradiction, since 'man' and 'mankind' meant the same thing.)

6 *Diacope* – repetition of a phrase or word, but with other words in between. 'Magnificent, it really is magnificent – in fact, the most magnificent example I've ever seen', for instance.

7 *Dirimens copulatio* – 'But wait – there's more!' That's an example of dirimens copulatio; amplifying an argument by building the story in a 'not only, but also' kind of way. As *Columbo* would have put it, 'Just one more thing ...'.

   It can be very effective, if rather cheesy. Just flick to any of the TV shopping channels and see how they build the benefits of a product.

   The classic iteration is where they sell the benefits of a product, tell you its price, you think that's the whole deal, it sounds pretty good but you're not quite persuaded ... and then the 'But wait – there's more' moment arrives. They tell you that when you order today, you'll also get these three great accessories free. And free delivery. And a money-back guarantee. And a free unicorn.

   Find a subtle, sophisticated way to use the same principle.

8 *Exemplum* – just means an example. This book is full of examples and they're useful in most forms of business writing. Not only do they make your story more interesting, they can act as an illustration of your subject that's far more compelling, because you're making it real.

9 *Hyperbole* – refers to lyrical exaggeration. You can do it in a crude way, like, 'The most incredible sale the world has ever seen!'.

Or you can do it more subtly, for example with an exaggerated metaphor such as, 'We were so poor, my parents got married just for the rice'. Or just by choosing descriptions that are more evocative and aspirational. Be careful though – hyperbole can become 'overselling'.

Essential example

The other day I was on a plane and the in-flight meal menu referred to 'Homemade roast potatoes'. That didn't impress me, it just struck me as ridiculous. How can airplane food be homemade? In whose home? The pilot's?

That kind of overboard exaggeration just undermines your whole communication. And don't scatter meaningless adjectives everywhere. The occasional superlative can make a noun stand out, but if everything is 'fantastic', 'amazing', 'incredible' then those words lose all meaning.

10 *Hypophora* – asking a question and then answering it. Often useful since your audience is likely to have questions about your product/service/offer, so you can ask those questions yourself, then answer them. For example, 'Where can you get the new LX8? It's available from Amazon right now – just click on the link below'.

11 *Mis-saying* – changing a well-known or common expression slightly. For a watch ad, the headline 'There's no present like the time', reverses the 'There's no time like the present' expression.

Reverse words or just change one of the words to another that rhymes with it or sounds like it – the less you change the better, so there's more chance people will get the original reference.

12 *Onomatopoeia* – words that sound like the thing they describe.

It's an interesting idea, since we're talking about writing that is unlikely to be read aloud – yet the way your words would sound out loud can still make a difference.

Examples of onomatopoeia include buzz, zip, screech, whirr, crush, sizzle, crunch, grind, mangle, bang, pow, zap, fizz, burp, roar, growl, blip, click, whimper and echo. In fact, onomatopoeia has been the whole premise of an enduring ad campaign – Rice Krispies' *Snap, Crackle & Pop*.

13 *Oxymoron* – a contradiction in terms. The cheeky example often used is 'military intelligence'. You might use a contradiction in terms to make a comparison or draw attention to a particular point. 'Deafening silence' and 'open secret' are oxymoronic phrases.

14 *Parallelism* – this just means repeating your linguistic structure in a phrase, sentence or paragraph. It creates a poetic rhythm.

Here's an example from Cormac McCarthy's *The Road,* using parallelism to build up a sequence of actions with a sequence of nouns: 'He pulled the blue plastic tarp off of him and folded it and carried it out to the grocery cart and packed it and came back with their plates and some cornmeal cakes in a plastic bag and a plastic bottle of syrup.'

There are many ways you can use parallelism. For instance, instead of different numbers of adjectives before each noun, as in 'a book, a yellow pencil and a worn, dirty coat', you could consistently always have one, as in 'a heavy book, a yellow pencil and a dirty coat'.

15 *Personification* – is where you give something that isn't human (usually an object), human qualities (such as emotions, senses, actions). So it can be as simple as 'Our new car just *begs* to be driven' or 'Milk's *favourite* cookie' (a strapline for Oreos) or 'Give your radio *a reason to live*' (a headline for Virgin Radio).

Perhaps one of the reasons personification works well is that we're obsessed with ourselves, so anything that's made more human is brought closer to us, made more real for us.

16 *Pleonasm* – is to deliberately use word redundancy; normally we 'eschew surplusage' as Mark Twain wittily put it, but sometimes repeating an idea, perhaps with different words, can help. It can also create emphasis, as in: 'The latest version is faster, quicker, more rapid' (which is an example of pleonasm and of the power of threes too).

17 *Pun* – another one, like hyperbole, that tends to be overused, especially by more junior writers. But don't *write them off* (there's a pun for you) – in the right hands, they can be awesome.

Let's say there are good puns and bad puns. In fact, good puns I just call 'wordplay', as 'pun' tends to be used as a derogatory term.

Puns are supposed to be (mildly) humorous or witty, so they only suit subjects where that's appropriate. A pun is typically either:

a) deliberately using a word in a way where it could have more than one meaning, or

b) using a word that sounds similar to the word you'd normally expect in the phrase.

An example of the first type is by Charles Saatchi, who claimed his favourite of his own ads was for a haemorrhoid cream, with the headline 'How to lick your piles'.

The English language is replete with words that have two (or more) meanings, so there's always lots of opportunity for wordplay of this kind.

An example of the second type, one that still makes me laugh for its sheer charming silliness, was the name of a scuba shop I once saw on holiday – 'Scuba Dooby Doo'. But my personal favourite of all time was (allegedly) an ad by an Irish company called Sofa King. Their strapline? 'It's Sofa King Good.'

If you're doing a presentation, a clever pun on screen can have a moment's entertainment value.

18 *Rhetorical question* – unlike hypophora, this means asking a question which you *don't* then answer. 'Who doesn't love a rhetorical question?' That's one right there.

'Do you want thicker, shinier hair?' might be a useful example for a shampoo ad, where the target audience would, internally, be saying yes. But to avoid people saying 'No' and dismissing your communication, consider *open questions* (ie, ones with lots of possible answers, like 'What does your favourite colour say about your personality?') instead of *closed questions* which can be answered 'yes' or 'no'.

19 *Rhyme* – it may seem trite, but rhyme is a great memory aid.

The Appliance of Science (Zanussi), O2 See what you can do, School fuel (Shreddies), Beanz Meanz Heinz or *The flavour of a Quaver is never known to waver*, for instance.

Most song lyrics rhyme of course – because it makes them more catchy and memorable. And similarly, don't we want our message to be memorable? For inspiration, there are plenty of online rhyming dictionary sites that let you type in a word and find all the words that rhyme with it (or use assonance for a near rhyme).

20 *Tautology* – means a phrase which contains an unnecessary repetition of meaning. 'Free gift' is the classic example. Because if something is a gift, then of course it's free. Otherwise it wouldn't be a gift. Yet it's often used because it just sounds stronger than simply saying 'free'.

'I saw it with my own eyes' is a kind of tautology too. Of course you saw it with your own eyes – you're not likely to see it with someone else's, are you?

Solutions not problems

Usually, it's more successful to talk about the solution rather than the problem. The example I mentioned in the flipchart exercise near the beginning of this book was anti-dandruff shampoo.

Sometimes they can do a 'Before and after' that shows first the problem (dandruff), then the solution (shiny, 'flake-free' hair). But in a space where they're leading with one main picture, it won't be of someone with dandruff. The picture will be of someone (a woman, usually) with glossy, beautiful hair. The outcome.

Straightforward testing, like for like, just shows that solutions outpull problems.

Essential example

Instead of 'Does our business suffer from . . .' (the problem) try 'Wouldn't it be great if our business . . .' (the solution).

Although, as always, it's not an absolute rule. It's often more creatively interesting to bring the problem to life rather than the solution, and a headline that leads on the problem certainly can work as it helps the target audience realise you're talking to them.

Nouns beat adjectives

Ever read any Ernest Hemingway? He wrote very sparingly. Not too much description. Very tight, focused writing.

The advantage for you of writing 'sparsely', by editing out some of your long-winded description, is that you can get the 'meat' across more quickly, cutting out the fat. It might also mean you let the reader do more of the imagining (which involves them more) rather than spoon-feeding them every detail.

But there's another point about nouns and adjectives – many ideas can be expressed as either a thing (a noun) or a characteristic of that thing (an adjective). And it's more powerful to express something as the noun, rather than the adjective. So if you really want to convey expertise, 'our expert staff' is not as potent as 'our staff of experts'.

Avoid talking about cost, buy, price, pay, spend, owe

All of the words above are negative. They're bad things to have to put in your writing. Why? Because they tell your reader that they're about to become poorer – which is always a bad thing. So minimise their use. In fact, try not to use them at all.

Instead, try something like 'Yours for just £XX.99'. Or for something pricey, try 'invest' instead of buy – because investing is a positive thing.

If possible, you should even avoid using 'spend' even in a non-financial sense, because it always has monetary connotations, even when you don't mean it that way. So don't say 'Spend two nights at this luxury hotel'. Instead, simply 'Enjoy two nights at this luxury hotel'.

Avoiding using the *word* 'price' doesn't mean you should avoid *having* the price. If you want someone to buy something, you must tell them how much it is. Just don't refer to it as a cost. There's evidence that people won't pursue buying something if they don't know how much it is – they're put off by the idea that they might have to enquire, then discover it's too expensive and feel embarrassed.

Quantify

Another point made in the flipchart exercise at the start was how people prefer hard numbers when describing amounts. Even if those numbers don't really mean much to them, the idea that you're being definitive is effective.

Lots of business writing is littered with phrases like 'some' or 'lots of people' or 'a number of' and so on – often simply because the writer didn't bother to find the actual numbers. Don't be like that. Instead of saying 'thousands of people have discovered ... ', use '15,000 satisfied customers have discovered ... '. Using a hard number just seems more powerful.

The exception is when the actual number is very disappointing. Clearly you wouldn't want to use it in that case. A final point to mention with quantities is how the language around them can affect the perception of that quantity.

Essential example

If you want to make an event seem more likely, say 'you may win' rather than 'you might win'. 'May' feels more likely than 'might'.

'Should you need to make a claim' is less likely than 'If you need to make a claim' (and therefore, if you want people to think they're unlikely to need to claim, better). 'As much as' means the same as 'up to' but sounds bigger, because it has 'much' in it.

So car insurance ads on TV shouldn't say 'Save up to 40%', they should say 'Save as much as 40%'. Actually, they should give the amount in pounds – it should be 'Save as much as £120'. And don't say '20% off', say 'Save 20%'.

Be active not passive

An active style means using 'you' and 'I' and 'they' so it's clear who we're talking about. A passive style loses this and creates distance between the topic and the reader. The only example I can think of where this is done deliberately is insurance booklets, where instead of saying 'if you have an accident' they'll say 'in the event of an accident'.

But, unless you're writing insurance booklets, avoid this passive style. Consider having plenty of 'you' and 'your' in your writing. You can see how often I use it throughout this book. In fact, I always go back through my writing to make sure I've used 'you' enough.

Essential tip

When you've written your first draft, check where the first occurrence of 'you' is. If it's too low down then consider sticking one in earlier. Even if that means contriving a new sentence that only exists to get the word 'you' in.

Keep it short (Or long)

Here are my two rules for writing length:

It should never be longer than the space it suits.

It should never be longer than you're capable of being interesting for.

The first point is one of simple logistics: you can't pour a quart into a pint pot. It's all very well thinking that every word you've written is sublime, but if you write more than the space you're working in can comfortably house, it'll look awkward and off-putting and no-one will read it. And don't settle for squeezing in the most it can possibly carry either: edit and trim your writing until it looks relaxed and elegant *in situ*.

The second point is more about the psychology of persuasion. It takes time to persuade someone to do something. So to a certain extent, the longer you can spend with your prospect, convincing them of your argument, the more chance you have of persuading them. To use a well-worn aphorism, 'The more you tell, the more you sell'.

That's why there are companies offering a free holiday, provided you agree to come to a three-hour talk on time-shares. Because they know that if they can present their case over three hours, you're much more likely to be persuaded.

It's why a good car salesman will strike up a conversation with you, explore what you're interested in and really go to town on why such-and-such a car is perfect for you. Because the more he talks, the more you're persuaded. All of which suggests that longer writing may be more effective than short, in whatever medium you're trying to persuade in.

Which begs the question, why is so much writing so short nowadays?

One reason may be that the writing just isn't very good. Because it's fine to write 100 lines (assuming you have the space) if all 100 lines are interesting. If you can write a compelling first sentence, your audience will read on to the second. If the second sentence is interesting, they'll read on to the third – and so on. You can have the call to action as the 100th line, provided the preceding 99 lines were interesting enough for the audience to get that far.

Another reason for shorter and shorter writing is the widespread belief that people are becoming more and more visual – they read less, they have shorter attention spans, they're too busy.

But people still like to be engaged. People still like what's interesting. In summary: your writing can be long. As long as it's interesting.

Avoid a woolly ramble

As Oscar Wilde said, 'I'm sorry this letter is so long – I didn't have time to make it shorter'.

Long writing is not the same thing as long-winded writing. As we'll look at in *Reviewing*, you should always edit. Ruthlessly. Saying things in twice as many words as needed is the same as being vague. It means your reader can't see the wood for the trees.

Always. Trim. The. Fat.

I always remember someone pointing out to me how particularly true that seems to be in emails: how rarely a 'storytelling' approach or any other kind of prevarication works. Get straight to the point in the subject line (providing it doesn't use words that fall foul of spam filters) or headline.

Don't get them disagreeing

A simple example is: don't describe something as 'unique' when the audience knows it's not. As soon as people start disagreeing with your words or questioning their credibility, they'll start questioning your product or service's credibility.

Similarly, it's easy to invent a world where everything is amazing, incredible, fantastic, blah blah blah – I've read plenty of business reports and proposals that were overblown with superlatives. It just makes the writer sound like a permanently enthralled Disney character, rather than make the subject sound any better.

Don't say incredible unless you can prove why, otherwise credibility starts to fall. In fact, the original meaning of 'incredible' is that it's 'not credible'. Same with 'fantastic' – if something is fantastical then it's not real.

Aim to get your reader on your side by saying things they can agree with. Get them nodding along saying 'yes' throughout and you'll make it harder for them to say 'no' when you ask for the sale. Again, it's what good salesmen do. They establish rapport, befriend you, agree with you and get you agreeing with them ... so they're harder to say no to when they try and flog you something.

Don't know it, feel it

The difference between rationally knowing and emotionally feeling something is huge – and vital for persuasive writing. How many times have you felt a swell of emotion when watching a film? Maybe shed a tear when a character dies ... even though you know the whole thing is made up?

Interestingly, emotions make something more memorable too. Your own memories – both good and bad – are generally the ones that were of emotional experiences.

Show not tell

People don't like to be told things as much as they like coming to the conclusion themselves (or thinking they did).

And when they do, it'll be a lot more powerfully believed, because it was their decision. You just have to nudge them in the right direction. For instance, say you're promoting going to see Irish horse racing. Do you tell people, 'Horse racing in Ireland is very exciting'? Does that make it *seem* exciting? Or do you say, 'The only sound louder than the thunder of hooves is the pound-pound-pounding of your heart'.

'Telling' is lazy. 'Showing' requires more thought (and sometimes more words) but is a lot more powerful, because it engages the audience more deeply.

Three's the magic number

Writing in threes is a charming, bewitching, seductive writing technique. See?

Lots of things work well, grouped in threes – from the Three Wise Men to the Three Musketeers. And look at any home interiors magazine – you'll see page after page of carefully 'dressed' rooms with a row of three vases, or three items the same colour or three pictures across the same wall.

For some reason, writing works well with threes too. It's a great way to add rhythm, add tone, and reinforce the point you're making. Look, I just did it again in that sentence – gave three benefits (and before that, I gave three examples – vases, items the same colour, pictures on a wall).

Consider the merits of a few threesomes in your business writing.

Tell them what you want

Obvious, I know. But many business documents seem rather shy when it gets to the crunch: asking the reader to do something. They blush and stutter and um and ah around the subject like a teenage boy asking the school hottie on a date.

The 'call to action' is a vital part of any communication. If you're writing to persuade someone to do something ... make sure you tell them what it is you want them to do – and be clear about it. If you've seen the film *Glengarry Glen Ross*, you'll know the brilliant scene with Alec Baldwin 'motivating' the sales team. 'A B C', he says, 'Always Be Closing'.

No ifs or buts. Literally, no 'ifs'. Don't say, 'So if you want ... ' as that shows doubt: it allows for the possibility that they might not want to do what you want them to. Don't entertain that idea – because if you show doubt then the prospect certainly will.

A great set of tube ads in 2010 were just type, no pictures, for dixons.co.uk. They were printed in the colours and fonts used by different high street stores; so there was one written to look a bit 'John Lewis', about John Lewis. The ad read:

> *Step into middle England's best loved department store, stroll through haberdashery to the audio visual department where an awfully well brought up young man will bend over backwards to find the right TV for you ...* ***then go to dixons.co.uk and buy it.***

It was a great, cheeky proposition – 'Use your favourite shops to find the things you like, then save money by buying them online at dixons.co.uk'. It had a great strapline under the logo too: *Dixons. The last place you want to go.*

Effectively *the whole ad* was a call to action. Never mind writing that sidesteps what it wants you to do, in these great ads *every word* was about what they wanted you to do.

Urgency

This relates to the previous point 'Tell them what you want'.

Explicit urgency isn't always appropriate in some forms of business writing. At least, not in the old-fashioned 'Hurry – everything must go' kind of way. But some *implicit* urgency in your communication is nearly always useful.

In a world of bewildering choice, non-stop distractions and information overload, a little direction can be well-received. So don't pussyfoot around too much. Make it clear what you want your reader to do, and create some sense of moving towards that action.

You don't have to get your reader to run, it's perfectly fine if they walk. Just so long as you get them moving. The idea behind a 'run' sense of urgency (eg, 'Reply now – offer ends soon') is to get your reader to act *while you've still got them persuaded*. Because the magical power of your prose will wear off over time.

To use *Glengarry Glen Ross* as an example again, in that film the salesmen are flogging real estate – and there's a cooling-off period where customers can cancel. Just as you can with lots of insurance products.

As I'm sure you know, cooling-off periods were introduced because people were being strongly sold to, buying the product then getting second thoughts a few days later, but were unable to back out of the commitment.

You don't want people to back out of the commitment your communication will have elicited from them. So create a sense of urgency to get them to act. Plus, a genuine feeling of urgency is exciting. It stirs the blood – and that, again, makes for a good communication.

Not all business writing suits this 'hard sell' approach. But just bear it in mind, and make sure you are creating a little bit of 'hustle', to get people moving. Even if it's just conveying a sense of 'Now that you know how good X is, why would you wait to enjoy X's benefits?'.

The tease

Let the reader know that you're going to tell them something interesting …

… in a moment.

Some people don't see the benefit of this approach – indeed, it can seem to contradict what I've said elsewhere, about being simple and leading with what's most interesting to your audience. But a tease can work well, providing the story which comes before 'the main event' is interesting in its own right. You may be explicit about the tease.

Essential example

'I'd like to tell you about how you can have the thick, healthy-looking hair you've always wanted. But first, let me ask you what you know about Eskimos.'

There you're effectively telling the audience what you're going to tell them 'in a moment'. However, you can have a more implicit tease.

Essential example

An article with the headline 'The thick, healthy-looking hair you've always wanted' then opens with the line 'Let me ask you something. Why is it that Eskimos always have such thick, shiny, healthy hair?'.

In that one, there's no 'But first', it just starts with a subject that doesn't seem to be directly related to the headline. But it's intriguing. And because you've promised me a benefit in the headline, I'll read on.

And when it turns out that the Eskimo story is pertinent to how I can have thick, healthy hair (remember, 'relevant abruption') then I'll feel rewarded for persevering. The 'reveal' will actually be more compelling than if you'd just come out with it straightaway. (I don't know about the Eskimos by the way, I just made that up. Maybe they do have great hair. Probably all that fish full of Omega 3 or something.)

AA Gill's restaurant reviews are a great example of the tease. 1000 words of review, the first 600 of which may have nothing to do with the restaurant, but you're waiting to see how it becomes relevant to the review.

Make it flow

Every business communication should flow from the first word to the last.

Something that can interrupt the flow is an abrupt change of subject. A simple way around it can be to add a few words that link the ideas together. A dab of word glue. Here are some words and phrases that can link sentences or paragraphs together to keep things flowing smoothly:

What's more,	In fact,
Surprisingly,	Without doubt,
Also,	There's more.
It gets better.	It gets worse.
Not only that,	And
Plus,	However,
Nonetheless,	Nevertheless,
One more thing.	So,
In addition,	Now
Furthermore,	Which means
But consider	Therefore,

Paint a picture

You can paint a convincing picture with a lot less than a thousand words.

These 'word pictures' are vital in your writing. They create a story in your audience's mind's eye. Gets them to 'picture the scene'.

If you can paint an effective picture – using evocative words, startling description and emotive language – you will involve the reader more. As soon as they're picturing the scene you've created, they're engaged with what you're saying.

Beginning with 'Imagine … ' is a simple trick to help write in a way that starts painting a picture.

Another way is to make the reader the subject, by writing it all in the second person.

Essential example

A piece I wrote for homeless charity Shelter began as if the reader (who we wanted to make a donation to their helpline) was someone who was about to become homeless: 'A bitterly cold winter's night. You're hungry. Tired. Terrified. All you own in a bag on your back. And your only hope, a free phone call. But . . . will we answer?'.

Reframe it

We all have our own way of thinking. What's obvious to you may be far from clear to me. What I find easy, you may find hard. We're all different, we process information in different ways and we have different experiences to draw upon and to evaluate what we're presented with.

So, what you consider to be the best way to explain a product, service, benefit, offer, idea or principle may not be the best one for me to absorb it. Try 'reframing it'. That is, expressing it in more than one way, through your writing. That way, there's more chance that you'll cover several different 'best ways' for different groups of people.

Back to the start

I've mentioned the reprise before – a simple trick where the end refers back to the beginning.

It's like a film that, in the final credits, reminds you of some of the best moments of the movie. 'Oh yeah,' you say, 'I loved it when that thing exploded.' Your communication may begin with a creative hook, a surprising line or an interesting idea. Therefore, near the end, refer back to that opening hook.

The Eskimo article about how to have thick, healthy-looking hair by taking this new supplement, for example. You begin by talking about Eskimos (who have lustrous, healthy hair and who eat lots of fish full of Omega 3). You then move on to how this new supplement is the best way for your body to get Omega 3, which will give the reader fabulous hair too.

You talk more about the supplement, the science behind it, how much it costs and where to buy it ... and you haven't mentioned Eskimos since perhaps the second line of your copy.

So now, at the end, refer back to them. Finish by saying how this new supplement is 'the second best way there is to have great hair – the best being to come from Eskimo lineage. Like the current Miss World, for instance.'

Breaking the law

And finally that old adage 'When you know the rules, you know when to break them'.

The preceding 25 *quick wins* are just tips. Not laws. Not objective, unassailable truths. And certainly not exhaustive.

They interact with each other, they depend on the circumstances, the timing, the medium, the audience. That's why, even in business, great writing is an art, not a science.

Use these tips as a starting point, not a destination. Add new ones. Try new things, new approaches, find new truths. Audiences continue to become more sophisticated. Media continue to fragment. Language continues to evolve. And so should your writing.

14

Essential Do Checklist

I hope you'll agree: we've covered a lot of ground here. Time to catch our breath for a moment and have a checklist reminder of what we've achieved.

1 We began by covering the basics of punctuation, grammar, usage and typography – side-stepping the mistakes many business writers often make and returning to the constant theme of simplicity.

2 Then we examined style and structure. Techniques for cultivating an appropriate, compelling tone of voice such as 'personable adjective' and the *Song Method* as a more contemporary, organic way to approach structure than the old-fashioned, pre-determined headings route.

3 After that, we took a tour of the fundamentals behind crafting your draft – what to pay particular attention to when composing your headlines for instance, and how to tie visual aids into the overall communication.

4 Then we were into winning our audience's heart and minds, with what I hope you'll find is a useful collection of 'universal motivations' for your audience, as well as the mantra to keep answering their cry of 'WIIFM?'.

5 That naturally led to seven writing techniques to keep your audience engaged – such as using intrigue or 'water-cooler moments'. And from there, we explored one of my personal favourites – how to use psychology in your writing, with the

Divine Dozen psychological triggers to influence your reader, frequently on a subconscious level.

6 And finally, I brought together 25 'quick wins' – writing tricks from the world of advertising, some of which I previewed in the opening flipchart exercise right at the start of the book. A few minutes scanning those will often give you a simple way to pep up your prose.

So. That really is a lot to juggle at any one time, and getting the balance between all those elements is a challenge even an experienced writer can struggle with sometimes. But then, driving a car is an incredibly complicated activity too. Remembering how to use the foot pedals, gear lever and steering wheel together the right way. While looking ahead, checking your rear view mirror and your side mirrors. While looking at and interpreting all the road signs confronting you and avoiding all the bad drivers out there. All while nodding solemnly to your partner, jabbering away in the passenger seat.

Yet you manage that, no trouble. Often with the radio playing at the same time. It becomes instinctive, through practice and experience. Powerful business writing is like that too. Read, learn and inwardly digest these *Doing* guidelines. Whichever you feel are most relevant, interesting or natural to you, work with. Get to know the others over time. And, over time, you won't need to wrestle with them so energetically. Just like driving, they'll become second nature and you won't even be consciously aware of all the clever stuff you're doing.

Excellent. Now that you've written something approaching a masterpiece, we'll conclude by looking at how you review your work to enhance it further, as well as how to review your own performance and determine learnings for the future. So that, each and every time, you achieve our original goal, set out back at the start of the book.

Powerful business writing that tells a simple, engaging, persuasive story.

Reviewing

Apparently, Mozart wrote music flawlessly, first time. No mistakes, no amends.

He just put quill to parchment . . . and sonata after symphony after concerto poured forth, each perfectly transcribed from his head, without so much as a demi-semi-quaver out of place. The rest of us, however, have to work our way up to perfection. Editing, polishing and re-writing. Again and again and again, 'til your computer's trash can is fuller than Elton John's head of hair.

Essential tip

Remember the words of Henry Ford. When told he'd just been lucky, he replied, 'Yes, I have been lucky. And I find the harder I work, the luckier I get.'

So, we're going to review your rough diamond . . . and polish it 'til it gleams. I also want to touch briefly on giving and getting feedback, before finishing with a look at how to keep improving your business writing over the rest of your career.

15

Perfect, prune & polish

Perfect

Your first instincts may be to review what you've done against your proposition, the objectives, whether or not it's absolutely right for the audience and so on.

However, I recommend you review it *in isolation* of all those things first. Just read it as a piece of great (or not) writing. Notice what you think works particularly well (or doesn't). How it seems overall. How it makes you *feel*. Whether or not anything leaps out as needing attention, *before* you appraise it against your objectives. That way, you're looking at the big picture and the overall effect of your communication, rather than diving straight into the detail.

After all, it's possible to write something that ticks all the individual boxes, but which somehow doesn't quite hang together.

Essential tip

Make your writing better overnight. By leaving it overnight.

Always give your communication a read with 'fresh eyes' by coming back to it the next day and reading it anew. Sometimes it's not the total number of hours you have to work on a piece that's important, it's having enough *elapsed* hours.

If you have one full day to write something, say eight hours in total, but it has to be done by the end of that day, the chances are that the finished piece won't be as good as if you'd had six hours spread over two days to work on it (say two sessions of three hours each).

Because in the hours between those two sessions, your brain will still be assimilating what you've done, thinking about it . . . and when you go back to it that second time, you'll read what you've done afresh and be able to improve it a notch or two quite easily.

Once you've reviewed what you've written simply as a powerful piece of business writing, check it against your proposition, of course. And particularly concentrate on the key areas of your communication we've already identified – the headings, the opening, the close, the way you use facts and figures, the tone of voice, the structure and so on.

Also use the chapters of this book as a quick reference; if your communication doesn't feel as persuasive as you'd hoped, go rifle through the psychological triggers in Chapter 12 and see what trick you're missing. Or if it seems to be fighting the medium, go get a refresher in Chapter 4.

This is a good time to ask someone else to read it too. Someone whose opinion you respect and who will be honest. Give them the background to it and the objective – then they can give you a true appraisal of whether or not they think it's on the money.

Either way, by asking someone else to read your work and by re-reading it yourself with fresh eyes you can tinker with what you've written to help it realise your original objective more effectively.

Prune

And another thing: you've written too many words. You verbose chatterbox.

Whenever you review your work you should be looking for opportunities to cut out what you don't need, finding ways to make everything clearer, sharper and shorter so your best lines aren't buried among weeds. Partly it's a matter of finding dead-ends; half-formed ideas that distract from the main thought and can be done away with. And partly it's a matter of tightening up every single element.

Make your opening sentence 20 per cent shorter. Lose a waffly paragraph. Re-write one long sentence with three commas into two sentences with no commas. Find a shorter word than 'remonstrate'. Get to the action faster. Eradicate any repetition. Lose the bit you love but everybody else says doesn't work.

Incidentally, how many times can you use the word 'and' consecutively in a sentence, so it still makes perfect sense and doesn't need any pruning? How about five?

> *Man walks into the Rose and Crown pub. 'I like your new sign', he tells the landlord, 'but there's a mistake.'*
>
> *'Really? What's that?' asks the landlord.*
>
> *'The signwriter hasn't separated the words', the man replies. 'So there's no gap between Rose and and and and and Crown.'*

Genius.

Essential example 1

Sometimes I'll edit copy other people have written – say, the Managing Director putting a new business proposal together.

He'll write: 'Current messaging in the sector is bland and inconsistent and does little to tackle some of what we believe to be the barriers which prevent many people from acting. Much of this advertising also appears aggressive and "selly" rather than nurturing or supportive, which is perhaps surprising given the sensitivities of the subject of personal debt.'

I'll edit that to become: 'Most personal debt advertising is either bland or aggressive. They lack the supportive tone that will engage your audience and they don't address the four reasons that stop people from acting.'

Now it's a third shorter. It's got a more active voice. Shorter sentences. It gets to the point faster and in a clearer structure. And it's more specific, changing 'the barriers' to 'the four reasons' – and it moves this insight to the end of the para, so you're interested in reading the next para to discover what those four reasons are.

Of course, it's easy to be ruthless with someone else's copy. You've got to develop the discipline to be ruthless with your own.

Essential tip

Remember, make sure your writing isn't littered with flowery adjectives and adverbs. Go through it and start cutting them out; they're making your copy look amateurish and you as if you're trying too hard.

Essential example 2

More 'real life' business writing I've seen – 'The previous programme was based around issuing a standard welcome pack prior to then being sent standard appeals which were predefined across the calendar. Through realigning all historic communication and their results from a calendar-driven sequence into individual timelines, X were able to isolate the optimum time to communication about each required action.'

See how you can simplify those two sentences and prune out the dead wood to make them communicate more powerfully:

'Previously, everyone received the same welcome pack and then standard appeals at set times throughout the year. Instead, we tailored the welcome pack and changed the programme to send appeals to individuals according to when our results analysis showed was the best time to contact them to get a response.'

I'd say it had been effectively pruned – even though, in this case, it's only four words shorter.

Now, sometimes good editing can make things longer. Because you're pruning to make things simpler and more powerful. And sometimes that can take a few more words.

Essential example 3

Another Managing Director proposal – 'This proposal has been developed following an initial exploratory meeting on 29th March. It outlines how we recommend approaching and quantifying the most suitable target audience for an integrated campaign, what propositions work for these groups and creatively how they are best expressed.

The assumptions and costs provided within this document are based upon the initial discussion only and thus could vary slightly upon further exploration.'

I edited that to: 'This proposal shows how we can help you identify, reach and recruit the most valuable audience groups.

We had a very stimulating session together on 29th March – and we've used that meeting as a springboard. To make the most of the tremendous opportunities you have and to develop the most effective propositions into a creative, integrated campaign.

We've made some assumptions (including around cost) based on that meeting; obviously we can pin these down as we work together more closely.'

It's nearly 20 per cent longer. But reads 100 per cent better. And, unlike the original, it leads with a clear promise to the reader – a benefit that encourages them to read on.

Polish

Here's another saying for you: *good is a good start*. You've perfected the segments of your communication to meet the objectives more closely, suit the audience more clearly or take advantage of the strengths of the medium more fully. You've also edited and pruned what you've written until it's tighter than a botoxed actress's forehead.

But reviewing your writing is also an opportunity to take the key strengths of what you've written and make every element *more so*.

Essential tip

Take a lesson from author Elmore Leonard. He produced '10 rules for writing' which are worth looking up. The last one was 'If it sounds like writing, I rewrite it'.

To help with your polishing, I recommend you bring back SOPHIE, who we met all the way back in Chapter 5. She's extremely useful here to systematically read through your writing and ask how you can make your communication:

- *simpler*?
- more *original*?
- more *powerful*?
- more *honed*?
- more *intelligent*?
- more *emotive*?

Here, you're just looking to augment what you've done by turning up the volume on these six key descriptors of powerful business writing. Often the act of writing a business communication – whether internal or external, business to business or business to consumer – is an all-consuming exercise bedevilled by detail. This is your chance to reflect on the grander themes and push it from good to great. These six simple adjectives are often the way to achieve it

And finally: be brave.

Polishing is your last act before you hit 'send' or 'print' or whatever you're doing with it. If you're having trouble polishing what you've done then maybe – just maybe – it's because what you've done is spectacular and beyond improvement. Alternatively, it may be just unsalvageably bad.

It probably feels late in the day – but if when you're trying to polish your writing you realise it just isn't working, then start again afresh.

Essential tip

One new draft now, written quickly – but benefiting from all the thinking and practice you've already done – will probably give you a better result than trying to tinker with something that just doesn't seem to be cutting it.

16

Giving & getting feedback

There are probably times when you need to give someone feedback on their business writing. As well as times when someone will be giving you feedback on yours. I've been in both situations hundreds of times; here's what I've learned from the experience.

Giving feedback

Giving feedback is a weighty responsibility. But reading this book should help tell you what to look for in the writing you're reviewing. Then it's a matter of deciding how you're going to go through your thoughts with the writer.

That's my first tip, by the way, they're 'thoughts', not 'amends'. Telling someone you've 'got amends' suggests you have some changes which need to be made, full stop. No discussion. Which may get the other person's back up. Amends are a monologue. Thoughts are a dialogue.

Here are six things to think about.

Be kind.

It's easier to destroy than create.

Start your feedback by playing 'angel's advocate'. The opposite of devil's advocate, instead of looking for what's wrong, look for what's right. Be vocal about everything that's good.

Apparently, when Gordon Brown was Prime Minister, ministers gave him a 'sh*t sandwich'. Good news first, then the bad news in the middle, then a bit of good news to finish. It was supposed to make the bad news (of which there was always plenty) more palatable.

So be gentle. And start with the big picture, not a small detail. It doesn't do you any favours if your first comment is about an apostrophe rather than the concept or tone or overall strength of the piece.

Be mean.

While your approach should be thoughtful and even positive, your substance should be genuine, honest and challenging. Not feedback that will water work down or lose its focus. Instead challenge the author to make the work as good as it can be.

Be brave.

Take a deep breath: surprised by what you've read? Palms a bit sweaty? Good. Not every business communication should make you nervous, but some of it should. Being creative means doing something new. Doing something different. And different can be scary. Great writing should evoke a visceral response, not a shrug of the shoulders.

And don't over-analyse: the audience won't spend 20 minutes studying the headline, neither should you. Remember your first impression, your gut reaction to it.

Think bigger than the brief.

If a piece of writing is 'off brief', is it *automatically* wrong? The communication might be better than the original purpose. If you just look at whether something ticks every individual box, you may miss the bigger picture. First of all, ask yourself: is it a great piece of work?

Think about the audience.

Giving feedback shouldn't be about what you like, it should be about what you think will persuade the audience – so put yourself in their shoes. And remember, the work wasn't even designed for the audience to 'like' – it was designed for them to respond to.

Don't solve the problem.

Don't say, 'This sentence needs to change from this to this'. Get the author to solve the problem. If you or someone else amends the piece, it'll end up with multiple voices and it won't flow.

Essential tip

If you're reviewing someone's writing, remember this: sometimes the bravest, best thing you can do is not change anything. Often a communication is passed around for comments, and people write on it just to show they've seen it and had an input. Sometimes it's better to read a piece of business writing and say 'It may not be how I would have personally done it . . . but I recognise that it's on brief, well written and a strong piece of work. I'm going to leave it unscathed.'

Getting feedback

I've talked about emotion throughout this book. Well, getting your work critiqued is certainly emotional. But if someone's criticising your work, it doesn't necessarily mean your work's bad. It may just be that the person criticising it doesn't know what the hell they're talking about.

So try not to take it personally. It may feel like a criticism of you or your ability or judgement, but it isn't. It's just a person disagreeing with your choice of words. Anyway, here are some approaches that may be useful to you when you get comments on what you've crafted.

Are they right?

It's easy to be over-protective of what you've written. But the truth is:

- What you've written is one solution to the challenge. There are a million other ways to tackle it and many of them are likely to be at least as good as the one you've chosen.
- Being forced to go back and look at what you've done with fresh eyes will, though it's hard to admit, almost always make it better.
- The person making the amend may be right. What they're suggesting may be better – because you're not going to be 100 per cent right, 100 per cent of the time.

So have an open mind and consider the pros and cons as objectively as you can. Compare their comments with the proposition, the objectives, the brand. Explore the possibility that the amend – or at least the intention behind it – may be valid.

Are they wrong?

Ask for the problem, not the solution

Instead of making overall comments, people sometimes write literal changes – especially if they're looking at it electronically and they use 'track changes' on their software.

So instead, get them to express their concerns and find out what they feel isn't right. Echo it back to them, to ensure you both have the same understanding, then you can make the amend as sympathetically with the existing writing as possible.

Have a solution in mind, not just an argument

Just as they should tell you the problem not the solution, you should have an idea for what you could do instead or differently to what they're suggesting. Just saying 'No, that won't work' doesn't really help. Discussing what would work may.

Have expertise and experience, not an opinion

A debate can quickly become a group of people giving their views, with everyone's views considered equal, or with someone who's more senior having more clout even though they may know less about the subject. So back up your 'opinions' with some facts based on your experience and expertise. Try and make comments feel more objective and less subjective.

Lose the battle, win the war

Choose your battles. Sometimes, amends that you don't agree with will be forced upon you. See which ones you can live with. A bit of give and take will help your case with the ones you really don't agree with, rather than just saying *no* to all of them.

How to minimise amends

Severe amends are sometimes a consequence of misunderstandings. What you've written is either not what the reader was expecting, or they just didn't know what to expect. Both of which are easy to correct. For instance, have early discussions with key stakeholders on what you're planning. 'Tissue meetings', they're sometimes called.

Then, sell what you've written. Don't just pop it into a blind email or leave it in an in-tray. Build it up before you reveal it. Recap the brief and the proposition and the story behind how you approached it. And describe it all in a way which means that the work you've done is clearly the best solution to everything you've said. So you've got them persuaded before they've even read it.

At the end of the day

Two final points:

1. It's easy to get caught up in the heat of the moment and fight against an amend which, when all's said and done, may be pretty minor. Don't harm a business relationship over whether it should be a colon or semi-colon.
2. However, it's easy to go the other way and say yes to everything. Not only is this likely to ruin what you've done, it also conveys a lack of pride and confidence in your work. If you make every change without a murmur, people will start wondering if you know what you're doing.

Anyway, that's enough about feedback. At this point, you've created a finely crafted, perfectly honed piece of work. You've succeeded in creating *powerful business writing that tells a simple, engaging and persuasive story*. And that communication, presentation, project or campaign is done.

But what about next time? Is it going to take as much blood, sweat and beers every time you write something? I would hope not. Which means the very last thing for us to discuss is how to get better at business writing. So you can achieve your goal more easily next time, and influence your audience even more powerfully.

17

Getting better

There are lots of people who work hard to get 'pretty decent' at something ... and then take their foot off the gas. But have you ever noticed how the best people in their field continue to work hard and strive hard even *after* they've got to the top? Course you have – you're one of them.

Sometimes, if I've got a meeting somewhere distant, I have to get up early. And of course, there are fewer cars on the road at that time. But do you know what? The ones that I do see seem to be more expensive. You see a higher ratio of executive saloons at 6:30 am than you do at 8:30 am. Why is that? Because the best people *continue* to be the first in and last out of the office.

What's more, there are plenty of studies which demonstrate that, often, characteristics believed to be innate talent are actually determined by the amount of deliberate effort you put in over a prolonged period.

I remember hearing about a study of professional violinists. All extremely gifted musicians at the top of their game. And they were interviewed about everything that might have possibly influenced their talent. When they started playing. Who their teachers were. Whether their parents had been supportive. Who they were influenced by. How musical their families were.

Now, if their ability was purely down to innate talent, you wouldn't expect to find a relationship between these external factors and what separated the best of the best from the rest. But they did find a relationship: practice. It turned out that the very best violinists had simply put more hours of determined effort into becoming better. Each person had totted up hundreds of hours of practice beyond what the next tier of violinists had put in.

You've done exactly the same thing in your career, I'm sure. Put the hours in to rise to the top and continue enhancing your abilities. Well, the same application is true of writing. The more you train, the more you gain.

In fact, writing is one of those things that everyone in business does, but few people try to improve at. People treat it like breathing: it's just something you do without thinking about. Instead, think of it like a sport. You don't get better simply by doing it unthinkingly, you get better by a conscious act of *determined effort*.

You've read this book, of course, which is an excellent start. It means, if I've done my job properly, you're already better than you were. And better prepared than those people yet to read it. Just don't read it and discard it. Keep referring back to it and re-reading parts of it at least for the next three months. And spend the next six months really putting 'too much' time into your business communications. Because the determined effort you put in now will reap dividends for the rest of your career.

Essential tip

Here's one more cheesy aphorism for you, the last in the book I promise. 'The difference between *try* and *triumph* is a little *umph*.'

To finish, here are six things you can do to aid your 'determined effort' and continue to improve as a powerful business writer.

1 Break the habit

We all have established patterns to our writing, a preferred approach. Like an old, favourite jumper. It may have a few bobbles and be wearing thin at the elbows, but it's comfortable and reassuring and we wear it far more often than we should.

Challenge yourself every once in a while to do things differently. I heard a story about someone who took a different route to work every day for a year. The idea was to see new things each time and arrive at work stimulated by the fresh experience each day. Try and do things differently in your business writing too – like making yourself use one of the conceptual approaches you've never tried before, or starting writing halfway through the communication, or trying a structure you've never used before.

2 Be a magpie

Collect examples of good and bad writing. Everything you read, appraise for the quality of the writing. Notice what you think works well and what fares poorly. Great business writing is fairly rare, in my experience. If you find something you admire, file it and re-read it for inspiration. It's always a useful starting point to read some powerful business writing on an unrelated subject, just to get you 'in the zone'.

Also collect up great quotes, explanations, examples, anecdotes and so on. You'll notice I've peppered this book with the ones that I remember. Make a note of interesting stories you come across and you might be surprised how often they're useful to illustrate your point.

3 Be a jack of all trades

This book touches on typography, grammar, layout, psychology and more. Because you should learn about everything that might improve your business writing without necessarily being *about* writing.

For instance, my first job out of uni was working in a telemarketing team. It taught me a lot about how to grab someone's attention and how to best phrase an offer to get the most people saying yes. My second job was in the print-buying team of a big marketing department. I learned loads about printing processes, paper stocks, production techniques, formats and so on. It's been useful ever since.

Consider how you can bring your unique experience and interests into what you write, to create a distinctive, accomplished voice. Explore the world far beyond your narrow business field and you'll find all sorts of ways to add life to your writing.

4 Be demanding

Of yourself: of course. But be demanding of the people who are involved with what you're crafting too. Persuade people to work hard to get the information you need. To help you develop a strong layout. To source great imagery. To find out more about your audience so you can really tailor your writing. Let them know that it matters to you, how good a job you do as a team.

Most business communications are as a result of several people's contributions – you're doing your damnedest to make it as powerful as possible, so don't have it let down by someone else not pulling their weight.

5 Six-month review

Keep your best business documents – the ones you're most proud of – and look at them again six months on. Are they still as perfect as you remember them? If not, then that's a good sign – it means your standards have improved and what was the pinnacle six months ago no longer is. But if every time you look back, what you've written is of the same standard as what you're writing now, then you're not improving. Which can only be down to one thing: you're not trying hard enough.

6 Enjoy it

Yes, predictable I know. But enjoying it *is* important. As someone wise once said, 'When you enjoy what you do, nothing is work'. Powerful business writing should be a pleasure to read. And as you get better at it, you may just find it becomes a pleasure to write.

18

Essential review checklist

And finally, here's a quick recap of *Reviewing*. First, augmenting what you've written:

1. Read it afresh, without referring to the background or objectives or audience and just get a feel for whether or not it falls into the category of powerful business writing.
2. Then check it against your original objectives, the proposition, the audience and the medium. Pay particular attention to the most critical segments of your pieces. Make some decisions about what needs to change as a result.
3. Take this final opportunity to cut off the flab and leave only the meat, and check what you've written against SOPHIE – simpler, more original, more powerful, more honed, more intelligent and more emotive.

Then we looked at giving and getting feedback; the importance of not sweating the small stuff but concentrating on the big themes and how to minimise changes.

Our last section looked at how to get better through application and determined effort, treating business writing like any other skill you want to develop. It means putting in extra effort on top of what you need to do the job. If a piece of business writing is going to take you 10 hours, actually put in 12 to 14 to allow yourself time to really push on and improve what you've done. The more you do it, the easier it will get, until you can achieve in 10 hours what used to take you 14.

And so here we are, at the end of our time together. I hope you've found it a useful experience.

Being able to 'write well' isn't a skill people put on their CV. There are no accredited exams you can take to prove your excellence. And when you're interviewing someone, you're probably unlikely to put their business writing skill to the test.

Yet it's clear that the ability to crystallise an idea, enthrall an audience and persuade senior decision makers is an enviable and invaluable one. It's an ability that powerful business writing gives you – and as you grow more comfortable with the tenets presented in this book, and through judicious practice, it's an ability you'll be able to demonstrate, whether you're writing a proposal, a board recommendation, an advert, an email or a presentation.

> *'Friends, Romans, countrymen, lend me your ears; I come to bury Caesar, not to praise him.'*

So begins one of the most famous speeches in all of Shakespeare's plays. This opening line to Antony's speech is a deliberate deception. When he takes the pulpit, the crowd is on the side of Brutus – the man who has murdered Caesar and justified the death to the crowd. Antony is telling them that's OK, he's only here for a burial, not to try and persuade them of Caesar's virtue through oration. Yet that's exactly what he then proceeds to do, praising Caesar with such clever rhetoric that he completely wins the hostile crowd over.

I'm not suggesting that following this book will make you as accomplished a linguist as Mark Antony (who had a pretty gifted speech writer, after all). But by being able to craft a simple, engaging and persuasive story, you will enjoy a significant advantage over those business people who haven't studied the craft of written persuasion.

I hope your new-found skills serve you well.

Index